Programmed Manual of College Physics

PROGRAMMED MANUAL OF COLLEGE PHYSICS

Jay Orear

Professor of Physics
Cornell University

JOHN WILEY & SONS, INC.
New York · London · Sydney · Toronto

Copyright © 1968 by John Wiley & Sons, Inc.

All rights reserved. No part of this book may
be reproduced by any means, nor transmitted,
nor translated into a machine language without
the written permission of the publisher.

10 9 8 7 6 5 4 3

Library of Congress Catalog Card Number: 68-8325
SBN 471 65685 2
Printed in the United States of America

PREFACE

SOCRATES: *"But if neither those who know, nor those who know they do not know, make mistakes, there remain only those who do not know and think that they know."*
ALCIBIADES: *"Yes, only those."*
SOCRATES: *"Then this is ignorance of the disgraceful sort which is mischievous."*

This programmed manual makes use of some of the latest developments in the rapidly growing field of programmed instruction. It has several advantages over conventional workbooks, textbooks, and other types of programmed workbooks, some of which are the following.

1. The above quotation from Plato warns us that one of the dangers in trying to learn is that there will be many who do not know but think that they know. A major purpose of this manual is to let them know just what it is they do not know. Experience with the first edition of this programmed manual has shown that one of its main advantages is to let the student know just what he does understand and just what he does not. As he works each problem, he can immediately check the correctness of his response against a list of answers.
2. Each student can work at his own pace and level of difficulty. Slower paced, easier material is automatically provided when a student runs into difficulty; deeper and more challenging material is occasionally provided for only the advanced and more talented students. Each student follows his own individualized path through the material. This path automatically adjusts itself to meet the student's needs.
3. In this manual students never get lost or stuck when trying to solve deep problems involving long chains of reasoning. Hence it is possible for students to work problems of greater length or difficulty than they could handle in conventional workbooks. A person should be able to do a reasonable job of mastering physics all on his own just by working through this programmed manual while studying his accompanying textbook.*
4. Preceding steps and their answers are often displayed on the same page and can be referred to in working the next step. This is a valuable feature for the programming of science and math. Most other teaching machines are incapable of this. Nearly all other programmed books and teaching machines attempt to be self-contained with the consequence that there is no convenient means provided for the student to refer back to earlier material. In science and math the current subject matter usually depends closely on earlier material and a means should be provided for ready display of earlier material in a concise, logical order. The means provided here for display and reference is the accompanying textbook—here it is the textbook rather than the program that is self-contained.

Ithaca, New York   JAY OREAR
August 1968

*This programmed manual was originally designed to accompany *Fundamental Physics* by Jay Orear, John Wiley & Sons, Inc., New York, 1967. The present format is based partly on suggestions of Dr. G. Vanderhaeghe.

INSTRUCTIONS

All empty answer boxes are to be filled in. Some of them already contain several choices for the answer, and in such cases determine which you think is correct and underline it. Complete answer sheets are located at the end of each chapter and a full set of duplicate answer sheets is at the end of the manual. Each answer sheet may be removed for ready reference. Note that the answers on a given answer sheet are not in order. Any particular answer may be quickly found, however, by referring to the item number to the right of the answer box.

In the sample frame shown there are two answer boxes which have item numbers 27 and 13 respectively. After filling in the first box check the 27th entry on the Chapter 1 answer sheet. You will find that it is *four*. Hence *four* is the correct answer. Next proceed to the second answer box and underline either the word *right* or the word *left*. Then check your choice by referring to item 13 on the Chapter 1 answer sheet, which turns out to be the word *left*. Hence *left* is the correct answer.

Sample Frame

1-1 To multiply by 10^{-4}, move the decimal point _____   **27**

places to the [right ; left] .   **13**

If incorrect go to 1-3; if correct go to 1-5

Note the instruction at the bottom of each frame which tells you where to go next. Occasionally the order of the frames will be deliberately scrambled. Do not work the frames in numerical order unless the instructions say to do so. Some of the frames (including the sample) involve what is called branching. This instruction means that if even one of your answers is wrong you must proceed to frame 1-3 (Chapter 1, frame 3); if all your answers in the frame were correct, go instead to frame 1-5 (Chapter 1, frame 5). In branching each student follows his own sequence of frames, which should be most suited to his particular needs; for example, in the Chapter 2 program there are 4096 different possible paths that could be followed. If each of these paths occurs with comparable probability, there is only about one chance in 4000 that your completed lesson for Chapter 2 and that of one of your friends should look the same. Certainly no two completed workbooks should be filled out in exactly the same way.

Most of the questions are made sufficiently easy so that an average student can usually get the correct answer (although there are occasional questions that nearly everyone is expected to miss). If you find yourself consistently getting more incorrect responses than correct, something is wrong. Too many incorrect answers probably mean that you are going too fast and are substituting guesswork for thinking. NEVER GUESS, WORK OUT EVERY PROBLEM CAREFULLY BEFORE MAKING YOUR RESPONSE.

The program in this manual is designed strictly for self-study. Hence the advantages of this programmed manual will be lost and your time wasted if you work it together with one or more students.

All answer boxes in your path are to be completed. Some questions such as the sample frame will require more than one response.

CONTENTS

CHAPTER 1. Introduction, 1

CHAPTER 2. Kinematics, 9

CHAPTER 3. Dynamics, 25

CHAPTER 4. Gravitation, 41

CHAPTER 5. Angular Momentum and Energy, 53

CHAPTER 6. The Kinetic Theory, 73

CHAPTER 7. Electrostatics, 85

CHAPTER 8. Electromagnetism, 103

CHAPTER 9. Electrical Applications, 127

CHAPTER 10. Wave Motion and Light, 143

CHAPTER 11. Relativity, 159

CHAPTER 12. Quantum Theory, 181

CHAPTER 13. Atomic Theory, 195

CHAPTER 14. The Structure of Matter, 209

CHAPTER 15. Nuclear Physics, 221

CHAPTER 16. Particle Physics, 231

Programmed Manual of College Physics

Chapter 1
INTRODUCTION

1-1 This chapter is a quick review of mathematics pertinent to the physics here. This short program should be adequate for those who have mastered the first two years of high school math. However, the program is not capable of teaching two years of math to those who have never really learned it in the first place.

Before presenting the problems we shall state without proof the general rules for handling exponents:

To multiply, add exponents: $A^m \times A^n = A^{m+n}$

To divide, subtract exponents: $\dfrac{A^m}{A^n} = A^{m-n}$

To raise to the Nth power, multiply exponent by N: $(A^m)^N = A^{Nm}$

To take the Nth root, divide exponent by N: Nth root of $A^m = (A)^{\frac{m}{N}}$

Go to 1-3.

1-2 You did not follow instructions. There was no way for you to arrive at this frame if you had followed instructions. In general the frames in this program are not in consecutive order. The correct order is determined by carefully following the instructions at the end of each frame.

Go back to 1-1.

1-3
(from 1-1)

One fifth of 10^{-5} is [10^{-1} ; 2×10^{-6} ; 5×10^{-5} ; 2^{-5} ; 2^{-1}].   4

If incorrect go to 1-18 ; if correct go to 1-11.
(This means if your choice was incorrect go to frame 18. If it was correct go to frame 11.

1-4
(from 1-23)

One fourth of 10^{-4} is [0.025 ; 0.0025 ; 0.00025 ; 0.000025].   14

If incorrect go to 1-19 ; if correct go to 1-5.

1-5
(from 1-4 or 1-19)

10^{-15} times 10^{-3} is _____ .   6

If incorrect go to 1-9 ; if correct go to 1-6.

1-6
(from 1-5 or 1-9)

10^{-15} divided by 10^{-3} is [5 ; 10^{-18} ; 10^5 ; none of these].   16

Go to 1-13.

1-7
(from 1-13)

$(10^{-8} \div 10^{4}) \times \dfrac{1}{10^6} = \left[\dfrac{18^{-8}}{10^{-4} \times 10^6} \; ; \; \dfrac{10^{-8}}{10^{10}} \; ; \; \dfrac{10^{-4}}{10^6} \; ; \; \text{none of these} \right]$.   2

$1 \div 10^4 =$ [10^4 ; $1/10^{-4}$; $1/10^4$; none of these].   12

$\dfrac{10^{-8}}{10^4 \times 10^6} =$ [10^2 ; 10^{-2} ; 10^{-18} ; none of these].   17

Go to 1-10.

1-8
(from 1-18, 1-27, or 1-12)

One fourth of 10^{-4} is [$(2.5)^{-4}$; $.25 \times 10^{-4}$; 4×10^{-4}].   19

Go to 1-23.

1-9
(from 1-5)

When multiplying powers of 10 the rule is to add the exponents. The general rule is
$$10^a \times 10^b = 10^{a+b}$$

The reciprocal of 10^{-3} is [$10^{-\frac{1}{3}}$; $10^{\frac{1}{3}}$; 10^3 ; none of these].   1

Go to 1-6.

1-10
(from 1-13 or 1-7)

If $A^{\frac{1}{3}} \times A^{\frac{1}{3}} \times A^m = A$, then the quantity m is _____.   21

$A^{\frac{1}{3}}$ is the [square root of A ; cube root of A ; neither].   23

Go to 1-14.

1-11
(from 1-3)

10^{-12} divided by 10^{-3} is _____ .   32

If incorrect go to 1-18 ; if correct go to 1-12.

1-12
(from 1-11)

$(8)^{\frac{1}{3}} =$ [2.67 ; $2\sqrt{2}$; 2 ; none of these].   18

If incorrect go to 1-8 ; if correct go to 1-14.

1-13
(from 1-6)

The rule for dividing powers of 10 is $10^a \div 10^b = 10^{a-b}$.

Hence $\dfrac{10^{-15}}{10^{-3}} = 10^{-15-(-3)} = 10^{-12}$.

Now try this one: $(10^{-8} \div 10^4) \times \dfrac{1}{10^6}$ is [10^{-2} ; 10^2 ; 10^{-10} ; 10^{-18} ; none of these].   24

If incorrect go to 1-7 ; if correct go to 1-10.

1-14 The reciprocal of *(A + B)* is $\left[\ \frac{1}{A}+\frac{1}{B}\ ;\ \frac{1}{A+B}\ ;\ -(A+B)\ ;\ \text{none of these}\ \right]$.   15
(from 1-12 or 1-10)

Go to 1-15.

1-15 $(8)^{\frac{2}{3}} = $ _____ .   8
(from 1-14)

Go to 1-24.

1-16 The square root of 9^4 is $[\ 9^2\ ;\ (4.5)^4\ ;\ 3^2\ ;\ \text{none of these}\ ]$.   25
(from 1-24 or 1-20)

If incorrect go to 1-25 ; if correct go to 1-21.

1-17 The square root of $9^4 = 9^{\frac{4}{2}} = \left(9^{\frac{1}{2}}\right)^4 = (3)^4$. This is of course also equal to 9^2.
(from 1-25)

Go to 1-21.

1-18 One fifth of 10^{-10} is $\left[\ 10^{-10} \div \frac{1}{5}\ ;\ 10^{-10} \times 2\ ;\ 10^{-10} \times \frac{1}{5}\ ;\ \text{none of these}\ \right]$.   3
(from 1-3)

If incorrect go to 1-27 ; if correct go to 1-8.

1-19 $10^{-4} = [\ 0.001\ ;\ .00001\ ;\ .00025\ ;\ \text{none of these}\ ]$.   20
(from 1-4) $\frac{1}{4} = [\ .25\ ;\ 2.5\ ;\ 1/.25\ ]$.   26

To multiply this by 10^{-4} we move the decimal point _____ places to the   9

[right ; left].   22

Go to 1-5.

1-20 $\frac{1}{A}+\frac{1}{B}=\frac{B}{AB}+\frac{A}{BA}=\frac{A+B}{AB}$. The reciprocal of this is $\frac{AB}{A+B}$.
(from
1-24)
$\frac{1}{A}-\frac{1}{B}=\left[\frac{B+A}{AB} \; ; \; \frac{B-A}{AB} \; ; \; \frac{A-B}{AB}\right]$

33

Go to 1-16.

1-21 The cube root of 9^6 is [9^3 ; 3^6 ; 9^2 ; none of these] .
(from
1-16,
1-25, or
1-17)

7

Go to 1-28.

1-22 In the triangle to the right, side A = _____ cm.
(from Side B = _____ cm.
1-28)

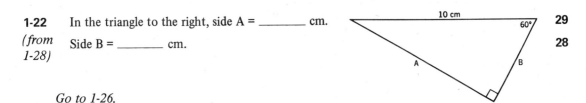

29

28

Go to 1-26.

1-23 One fourth of 10^{-4} is $\frac{1}{4} \times 10^{-4}$ and this can be written as
(from
1-8) [2.5×10^{-3} ; 2.5×10^{-5} ; 10^{-1} ; none of these] .

30

Go to 1-4.

1-24 By definition the reciprocal of A is $\frac{1}{A}$. The reciprocal of the quantity $\frac{1}{A}+\frac{1}{B}$ is
(from
1-15) then $\left[(A+B) \; ; \; \frac{AB}{A+B} \; ; \; \frac{A+B}{AB} \; ; \; \text{none of these}\right]$.

10

If incorrect go to 1-20 ; if correct go to 1-16.

1-25 Since the rule is that the nth root of $A = A^{\frac{1}{n}}$, the square root of (9^4) is
(from
1-16) $(9^4)^{\frac{1}{2}} = 9^{\frac{4}{2}} = 9^2$. The square root of 9^4 is 3^4 [true ; false].

31

If incorrect go to 1-17 ; if correct go to 1-21.

5

1-26
(from 1-22)
List all angles that are equal to angle 1.

_____

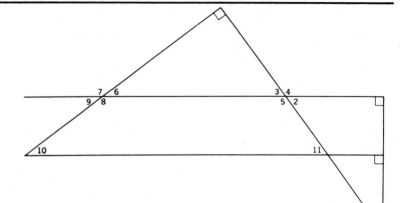

Go to 1-29.

1-27
(from 1-18)
The word "of" always means to multiply, not to divide. For example:

One half *of* 10 = $\frac{1}{2} \times 10$, it does not equal $10 \div \frac{1}{2}$.

$\frac{1}{2} \times 10 = 5$, but $10 \div \frac{1}{2} = 20$.

Obviously one half of 10 is 5 and not 20.

Go to 1-8.

1-28
(from 1-21)
In the preceding problem the cube root of $9^6 = (9^6)^{\frac{1}{3}} = 9^{\frac{6}{3}} = 9^2$. The cube of $\sqrt{x}$ is
[x^3 ; $x^{1.5}$; $x^{\frac{2}{3}}$; none of these] .

Go to 1-22.

1-29
(from 1-26)
This is a short, simple quiz covering some of the material in Chapter 1. Fill in the blanks. Answers to this quiz are not given in this manual.

1. The square root of 10^6 is _____ .
2. One fifth of 10^{-10} is _____ $\times 10^{-10}$.
3. One fifth of 10^{-10} is _____ $\times 10^{-11}$.
4. The reciprocal of $\left(\frac{1}{A} + \frac{1}{B}\right)$ is $\frac{1}{\frac{1}{A} + \frac{1}{B}}$ True _____
   False _____
5. $\frac{10^5}{10^7} =$ _____ .
6. $\frac{10^{-5}}{10^{-7}} =$ _____ .
7. $(\sqrt{x})^3 = \sqrt{x^3}$ True _____ .
   False _____ .

Chapter 1 Answer Sheet

Item Number	Answer
1	10^3
2	$\dfrac{10^{-8}}{10^{10}}$
3	$10^{-10} \times \dfrac{1}{5}$
4	2×10^{-6}
5	6, 9, 10
6	10^{-18}
7	9^2
8	4
9	4
10	$\dfrac{AB}{A+B}$
11	$X^{1.5}$
12	$\dfrac{1}{10^4}$
13	left
14	0.000025
15	$\dfrac{1}{A+B}$
16	none
17	10^{-18}
18	2
19	$.25 \times 10^{-4}$
20	none
21	$\dfrac{1}{3}$
22	left
23	cube root of A
24	10^{-18}
25	9^2
26	.25
27	4
28	5
29	$5\sqrt{3}$
30	2.5×10^{-5}
31	true
32	10^{-9}
33	$\dfrac{B-A}{AB}$

Chapter 2
KINEMATICS

2-1 A car starting from rest speeds up uniformly in time (constant acceleration) and reaches a speed of 30 mph in 1 minute. Of the curves shown below, Curve _____ best describes this situation. The average velocity of the car during this minute is _____ mph. (Hint: what is the average height of the curve?)

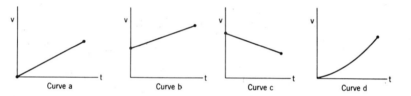

Now suppose that the car speeds up uniformly in time from 30 to 60 mph. Curve _____ best describes this situation. The average velocity of the car during this time is _____ mph.

In both cases the distance covered by the car increased uniformly with time.

[*true ; false*]

Of the curves shown below, Curve _____ best describes how the distance varies with time during the first minute as the car starts up from the rest. Curve _____ best describes the situation when the car is speeding up uniformly from 30 to 60 mph.

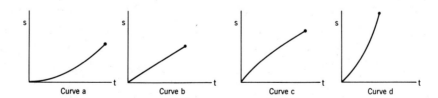

Go to 2-2.

2-2
(from 2-1)

A car starting from rest spends 1 minute getting up to 40 mph and then another minute getting up to 60 mph shown in the curve to the right. The average velocity during the first two minutes is _____ mph.

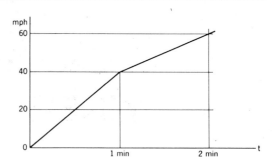

6

If incorrect go to 2-7 ; if correct go to 2-3.

2-3
(from 2-2 or 2-7)

In the previous problem the average velocity for the first minute alone is 20 mph. In order to find how far the car traveled in the first minute we use the formula

[$s = \bar{v}t$; $s = vt$ (where v = 40 mph) ; none of these] .   11

The result is $s =$ [1/3 mi ; 2/3 mi ; 20 mi] .   14

Using the same procedure we find that the car traveled _____ during the second minute.   7

Hence the total distance traveled in the two minutes is _____ .   12

One method for finding the average velocity of the car for the first two minutes is to use the formula $\bar{v} = \frac{s}{t}$ where $s =$ _____ mi and $t =$ _____ hr.   10, 16

The result is $\bar{v} = \frac{7/6}{2/60}$ mph = 35 mph.

Go to 2-4.

2-4
(from 2-3)

The car in the above problem reaches a velocity of 60 mph or 88/ft sec in 2 minutes. Hence its average acceleration during the first two minutes is _____ ft/sec².   13

If incorrect go to 2-11 ; if correct go to 2-5.

2-5
(from 2-4 or 2-11)

A ball thrown up in the air has zero velocity when it reaches its highest point. At that instant its acceleration is _____ .   18

If incorrect go to 2-12 ; if correct go to 2-6.

2-6
(from 2-5 or 2-12)

The acceleration of a 1-kg mass sitting on a table is _____ m/sec².   15

Go to 2-13.

2-7
(from 2-2)

In the previous problem, the average velocity for the first minute only is _____ mph,   20
and the average velocity during the second minute only is _____ mph. (Refer to the   22
figure on page 10.)

The average during the two minutes together is merely the average of the two above answers, which is _____ mph.   17

Go to 2-3.

2-8
(from 2-13)

If you missed the previous problem you are having trouble with the basic definition of acceleration. After calculating the accelerations in the earlier car problem, you should not have missed this easier problem. This last problem was a direct application of the definition of acceleration, $a = \dfrac{v - v_0}{t}$. $a = \dfrac{0 - 88 \text{ ft/sec}}{2 \text{ sec}} = -44 \text{ ft/sec}^2$.

A car starting from rest reaches a velocity of 60 mph (88 ft/sec) in 2 minutes. Its average acceleration is _____ ft/sec².   25

Go to 2-9.

2-9
(from 2-8 or 2-13)

The formula $\bar{v} = \dfrac{s}{t}$ is not valid when the acceleration is changing with time.

[*true* ; *false*]   19

If incorrect go to 2-14 ; if correct go to 2-15.

2-10
(from 2-15)

With respect to the water a ferryboat is sailing 30° west of north at 12 mph. The stream is flowing 6 mph due east. As viewed from the shore, the direction of travel for the ferryboat is _____ and its velocity is _____ mph.

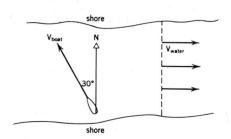

27

21

If incorrect go to 2-16 ; if correct go to 2-17.

2-11
(from 2-4)

Just as average velocity is given by the equation $\bar{v} = \dfrac{s - s_0}{t}$, so is average acceleration given by $\bar{a} = \dfrac{v - v_0}{t}$. In frame 2-4 the quantity $(v - v_0)$ is _____ and t is _____. The final results is $\bar{a} = $ _____ ft/sec².

Another way of working from 2-4 would be to calculate the acceleration for each minute and then average the two values. For the first minute the acceleration is _____ ft/sec².

For the second minute the acceleration is _____ ft/sec².

The average of the two values is _____ ft/sec. .

23

26, 28

30

24

33

Go to 2-5.

2-12
(from 2-5)

The acceleration due to gravity is constant and always acting on the ball, no matter what its velocity. The acceleration due to gravity causes the velocity to be constantly decreasing at a rate of 9.8 m/sec². Even when the ball reaches its highest point, its velocity is constantly decreasing at a rate of 9.8 m/sec².

Question: At its halfway point the acceleration of a projectile fired at a 45° angle is zero.
[*true ; false*]

29

Go to 2-6.

2-13
(from 2-6)
When an object is sitting on a table, not only is the velocity zero but it remains zero (no change of velocity).

Question: A car moving at 88 ft/sec (60 mph) comes to a dead stop in 2 seconds. Its average acceleration during those two seconds is _____ ft/sec². **35**

If incorrect go to 2-8 ; if correct go to 2-9.

2-14
(from 2-9)
Remember, from the basic definition of a weighted average, that

$$\bar{v} = \frac{v_1 t_1 + v_2 t_2 + \cdots + v_n t_n}{t_1 + t_2 + \cdots + t_n} = \frac{s_1 + s_2 + \cdots + s_n}{t} = \frac{s_{\text{total}}}{t}.$$

This is true no matter how v is changing with time.

Go to 2-15.

2-15
(from 2-9 or 2-14)
For an object starting from rest and speeding up uniformly in time to a velocity v, we have seen that $\bar{v} = \frac{1}{2}v$; hence $\frac{s}{t} = \frac{1}{2}v$ or $v = 2\frac{s}{t}$. But the usual equation for velocity is $v = \frac{s}{t}$ and not $2\frac{s}{t}$. This discrepancy is resolved by noting that these equations apply to different situations. Consider the following three situations:

a. Traveling with uniform velocity.

b. Starting from rest and traveling with uniformly increasing velocity.

c. The velocity may be changing with time in a nonuniform way.

The equation $v = \frac{s}{t}$ applies only to situation _____ . **37**

The equation $v = 2\frac{s}{t}$ applied only to situation _____ . **31**

The equation $\bar{v} = \frac{s}{t}$ applies to situations _____ . (More than one choice may be correct.) **34**

Go to 2-10.

2-16
(from 2-20)

The previous problem involves the vector addition of velocities.

To the velocity of the boat $\longrightarrow$ 6 mph. one must add the velocity of the water

The vector sum is v

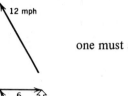

Note that this is a 30°-60° triangle. Hence the long side v is $\sqrt{3}$ times the short side v_{water} with the result that $v = 6\sqrt{3}$ mph. We could also use the Pythagorean theorem:

$$v^2 = v^2_{boat} - v^2_{water}$$
$$v = \sqrt{12^2 - 6^2} \text{ mph}$$
$$= 6\sqrt{3} \text{ mph}$$
$$= 10.4 \text{ mph}$$

Hence we have the result that according to vector addition (12 + 6) can equal 10.4.

Go to 2-17.

2-17
(from 2-10 or 2-16)

The time for the ferryboat to make a round trip in the N-S direction as viewed from the shore is _____ compared to the time required to make the same round trip in the E-W direction. 40

The time for the round trip in the E-W direction is _____ hours if the total distance is 2 miles. 32

Go to 2-23.

2-18
(from 2-17 or 2-13)

A projectile that was fired at an angle of 45° has a velocity of 1000 ft/sec when at its highest point. The horizontal component of its initial velocity is _____ ft/sec.

The total initial velocity is $V_0 =$ _____ ft/sec.

36

41

Go to 2-19.

2-19
(from 2-18)

A man falling out of an airplane will accelerate in free-fall until air resistance causes him to reach a terminal velocity of about 140 mph or 200 ft/sec (at this velocity the force of air resistance on a human body is just strong enough to cancel out the force due to gravity). Assuming a constant acceleration of $g = 32$ ft/sec², how far must an object fall (starting from rest) for it to reach a velocity of 200 ft/sec? The most expedient equation for solving this problem is $\left[v = 2s/t \; ; \; s = \frac{1}{2}at^2 \; ; \; v^2 = 2as \; ; \; v = at \right]$. The solution is $s =$ _____ ft. Now suppose a man who fell many thousands of feet out of an airplane landed in a soft snow bank which slows him down with constant acceleration. Assume that he will survive if the acceleration is less than 30 g. His velocity upon entering the snow is, of course, 200 ft/sec. In order for the man to live through this experience the snow bank must be at least _____ ft thick.

45
38

43

Go to 2-20.

2-20
(from 2-19)

We shall now derive a formula for the range of a projectile in terms of its initial velocity V_0 and angle θ. Let T be the total time the projective is in flight. Then the range $R = [\; V_x T \; ; \; V_y T \; ; \; V_0 T \; ;$ none of these $]$ where V_x and V_y are the x and y components of V_0.

V_x has the value $\left[V_0 \cos\theta \; ; \; V_0 \sin\theta \; ; \; \dfrac{V_0}{\cos\theta} \; ; \; \dfrac{V_0}{\sin\theta} \right]$.

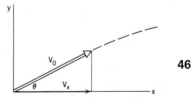

46

48

If incorrect go to 2-26 ; if correct go to 2-21.

2-21
(from 2-20 or 2-26)
In the above problem the range $R = \left[V_0 T \cos\theta \; ; \; V_0 T \sin\theta \; ; \; \dfrac{V_0 T}{\cos\theta} \; ; \; \dfrac{V_0 T}{\sin\theta} \right]$.   39

Let us now look at the vertical motion only. The time for the projectile to fall from its position of maximum height to the ground is _____. This time is related to the   42

vertical velocity V_y by the equation $V = at$. Application of this question gives the result

_____.   49

By systematically looking at the horizontal motion and vertical motion separately, we have obtained two simultaneous equations with the unknowns R and T. We can eliminate T by solving the last equation for T and substituting this expression for T into the previously obtained equation for R. This procedure gives $R = $ _____.   51

Go to 2-24.

2-22
(from 2-28)
The distance covered per revolution of the 4000 mi high satellite is _____ times the circumference of the earth.   44

Go to 2-30.

2-23
(from 2-17)
The total time T for the round trip is $T = s_1/v_1 + s_2/v_2$ where s_1 and s_2 are each _____ mile(s). When the boat is traveling east,   54
$v_1 = $ _____ mph. When traveling west,   47
$v_2 = $ _____ mph. Hence $T = $ _____ hr for the east-west round trip.   55
  50

Go to 2-18.

2-24
(from 2-21)
Those who know trigonometry will remember the double-angle relation: $\sin 2\theta = 2 \sin\theta \cos\theta$. If we make use of this formula, our result for the range of a projectile becomes $R = \dfrac{V_0^2}{g} \sin 2\theta$. The quantity $\sin 2\theta$ is maximum when (2θ) is _____.   52

Hence, to achieve maximum range, a cannon should be fired at an angle of _____.   57

The maximum range is $R_{max} = $ _____.   60

Go to 2-25.

2-25
(from 2-24)
Consider an earth satellite in a circular orbit. Then the magnitude of its velocity will be [constant ; changing with time] . 53

Its acceleration will then be zero. [*true* ; *false*] 62

The direction of its acceleration will be _____ . 56

Go to 2-27.

2-26
(from 2-20)
The definition of $\sin\theta = \dfrac{y}{z}$.

The definition of $\cos\theta = \dfrac{x}{z}$.

Then $x =$ _____ . 63

Go to 2-21.

2-27
(from 2-25)
Suppose the earth satellite is at a height of 4000 mi above the earth's surface. At this height the gravitational acceleration is $\dfrac{1}{4}$ the value at the earth's surface. Then $a =$ _____ is the acceleration of the satellite. The acceleration of the satellite is $\left[\dfrac{\frac{1}{4}v^2}{r} \; ; \; \dfrac{\frac{1}{2}v^2}{r} \; ; \; \dfrac{v^2}{r}\right]$ where r is the distance to the center of the earth (8000 mi). The acceleration is also $\left[\dfrac{\frac{1}{4}v^2}{R_e} \; ; \; \dfrac{\frac{1}{2}v^2}{R_e} \; ; \; \dfrac{v^2}{R_e} \; ; \; \dfrac{2v^2}{R_e}\right]$ where R_e is the radius of the earth (4000 mi). 58 66 64

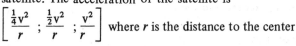

Let v_c be the velocity of a low-flying earth satellite. Then $\dfrac{v_c^2}{R} = g$. The velocity of our 4000 mi high satellite is _____ times that of a low-flying satellite. 59

If incorrect go to 2-29 ; if correct go to 2-28.

2-28
(from 2-27 or 2-29)
The period T of our 4000 mi high satellite is [1 ; $\sqrt{2}$; $\sqrt{8}$; 4 ; 8 ; none of these] times that of a low-flying satellite. 67

If incorrect go to 2-22 ; if correct go to 2-33.

2-29
(from 2-27)

For the satellite at a distance $r = 2R_e$ we have

1) $\dfrac{v^2}{(2R_e)} = \dfrac{1}{4}g$.

For the low-flying satellite we have

2) $\dfrac{v_c^2}{R_e} = g$.

Now divide eq. 1 by eq. 2. This gives

$\dfrac{v^2}{2v_c^2} = \underline{}$. 61

Or $v^2 = \underline{}\ v_c^2$. 69

Go to 2-28.

2-30
(from 2-22)

If the velocity is $\sqrt{2}$ times slower and the distance twice as great, then the time to cover this distance will be _____ times as long. Hint: $t = \dfrac{s}{v}$. 65

Go to 2-31.

2-31
(from 2-30)

In uniform circular motion the period of revolution is

$T = \left[\dfrac{R}{v}\ ;\ \dfrac{2\pi R}{v}\ ;\ \dfrac{R}{2\pi v}\ ;\ \text{none of these} \right]$. 72

Go to 2-35.

2-32
(frpm 2-33)

In uniform circular motion the velocity v is given by the formula $v = \dfrac{s}{T}$ where T is the period and $s = \underline{}$. 68

Go to 2-35.

2-33
(from 2-28)

In uniform circular motion the period of revolution is

$T = \left[\dfrac{R}{V}\ ;\ \dfrac{2\pi R}{V}\ ;\ \dfrac{R}{2\pi V}\ ;\ \text{none of these} \right]$. 82

If incorrect go to 2-32 ; if correct go to 2-34.

2-34
(from 2-33)
We want to derive a formula for the period of an earth satellite in circular orbit at a height h above the earth's surface. As we shall see in Chapter 4 the inverse square law of gravitation tells us that the acceleration due to gravity at height h is

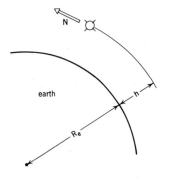

$$a = g \frac{R_e^2}{(R_e + h)^2}.$$

The net acceleration on the satellite is then

$$\left[\frac{4\pi^2}{T^2} R_e \ ; \ \frac{4\pi^2}{T^2} (R_e + h) \ ; \ g \ ; \ \frac{4\pi^2}{T^2} (R_e + h) + \frac{gR_e^2}{(R_e + h)^2} \right] \quad 85$$

Equate this to the above expression and solve for T:

$$T = \underline{\qquad\qquad}. \quad 84$$

The ratio of the squares of the periods of two earth satellites is

$$\frac{T_1^2}{T_2^2} = \underline{\qquad\qquad}. \quad 83$$

(This result is called Kepler's Third Law.)

The velocity of this satellite is $v = \frac{s}{T}$ where

$s = \underline{\qquad\qquad}$. In terms of v_c, R_e, and h, $v = \underline{\qquad\qquad}$ where 86
74
$v_c = \sqrt{gR_e}$ is the velocity of a low-flying earth satellite.

As air resistance causes the height of an earth satellite to decrease, the velocity will
[decrease ; increase ; remain the same] . 70

Go to 2-35.

2-35
(from 2-31, 2-32, or 2-34)

The number of revolutions in a time t is $n = \frac{t}{T}$ where T is the time for one revolution.

Let $f = \frac{n}{t}$ be the frequency or number of revolutions per second. Then $f = $ 73

$$\left[2\pi T \ ; \ \frac{2\pi}{T} \ ; \ \frac{1}{T} \ ; \ \frac{1}{2\pi T} \right] .$$

Go to 2-36.

2-36 The formula for centripetal acceleration $\dfrac{v^2}{r}$ can be expressed in terms of f and R.
(from 2-35) Then $a_c = \left[(2\pi f)^2 r \; ; \; \dfrac{4\pi^2}{f^2} r \; ; \; (2\pi f r)^2 \; ; \; \dfrac{4\pi^2 f^2}{r} \right]$. 77

If incorrect go to 2-40 ; if correct go to 2-37.

2-37 The units of $\dfrac{v^2}{g}$ where g is 980 cm/sec² are _____ . 71
(from 2-35 or 2-40)

Go to 2-39.

2-38 The sum of two vectors having magnitudes of 12 and 20 units could never have a magnitude less than 20 units [*true* ; *false*] . 75
(from 2-39)

Go to 2-41.

2-39
(from 2-37)

In this problem we wish to find out how the acceleration due to gravity decreases with the distance from the center of the earth. How can we measure the acceleration due to gravity at a far distance? Isaac Newton was the first to realize that the moon could be used for this purpose. The moon is an earth satellite at a distance of $R_m = 240{,}000$ mi $= 60\, R_e$. An object at a distance of 240,000 mi from the earth will fall toward the earth with the same gravitational acceleration as the moon. [*true* ; *false*] . The acceleration due to gravity at a distance of 240,000 mi will have the same value as the centripetal acceleration of the moon [*true* ; *false*] . The period of the moon is 27.3 days or 465 times that of a low-flying earth satellite. Let T_o and R_e be the period and orbit radius of a low-flying earth satellite. Then the acceleration of a low-flying earth satellite is $a_0 = g = \dfrac{4\pi^2}{T_o^2} R_e$. The acceleration of the moon is

$$a_m = \left[\; \dfrac{4\pi^2}{T_o^2} R_e^{\,2} \;;\; \dfrac{4\pi^2}{465 T_o^{\,2}} (60 R_e) \;;\; \dfrac{4\pi^2}{(465 T_o)^2} (60 R_e) \;;\; \text{none of these} \;\right].$$

79

80

76

The ratio of these two accelerations is $\dfrac{a_m}{g} = \dfrac{60}{(465)^2} = \dfrac{1}{3600}$ [*true* ; *false*] .

81

Note that the ratio of the squares of the two distances also happens to have this same value $\dfrac{R_o^2}{R_m^2} = \left(\dfrac{1}{60}\right)^2 = \dfrac{1}{3600}$. Hence, within the accuracy of our calculation the acceleration due to gravity appears to decrease at a rate inversely proportional to the square of the distance from the center of the earth. Isaac Newton performed this very same calculation on that famous day when he observed an apple fall from a tree.

Go to 2-38.

2-40
(from 2-36)

We can start from the expression for a_c in terms of T and r: $a_c = \dfrac{4\pi^2}{T^2} r$. Now substitute $\dfrac{1}{f}$ for T in this expression. The result is $a_c = $ _____ .

78

Go to 2-37.

2-41
(from 2-38)

Review quiz: We conclude with a short, simple quiz covering some of the material contained in the preceding questions. Use pencil.

(Answers are not contained in this Programmed Manual.)

1. A car travels at 20 mph for the first minute and 30 mph for the next two minutes. What is the average velocity during the first 3 minutes?

    $\bar{v} =$ _____

2. The acceleration time for a certain car to reach 60 mph (88 ft/sec) starting from rest is 22 sec. The average acceleration is

    $\bar{a} =$ _____

3. A boat is traveling at 5 mph due north with respect to the water, and the water is flowing 4 mph due west. With respect to the shore the boat is moving _____ mph in the _____ direction.

4. What is the acceleration due to gravity at a distance of 3 earth radii from the center of the earth?

    $a =$ _____ g.

5. If a satellite makes 6 complete revolutions per day, its period is

    $T =$ _____ hours.

6. All earth satellites in orbit are accelerating. True or false? _____

Chapter 2 Answer Sheet

Item Number	Answer	Item Number	Answer	Item Number	Answer
1	d	37	a	73	$\dfrac{1}{T}$
2	false	38	625		
3	a	39	$V_0 T \cos\theta$	74	$v_c \sqrt{\dfrac{R_e}{R_e + h}}$
4	45	40	shorter		
5	15	41	$1000\sqrt{2}$	75	false
6	35	42	$\dfrac{1}{2} T$	76	$\dfrac{4\pi^2}{(465 T_0)^2}(60 R_e)$
7	$\dfrac{5}{6}$ mi	43	21		
8	b	44	2	77	$(2\pi f)^2 r$
9	a	45	$v^2 = 2as$	78	$4\pi^2 f^2 r$
10	$\dfrac{7}{6}$	46	$V_x T$	79	true
11	$s = \bar{v} t$	47	18	80	true
12	$\dfrac{7}{6}$ mi	48	$V_0 \cos\theta$	81	true
		49	$V_0 \sin\theta = g\left(\dfrac{1}{2} T\right)$	82	$\dfrac{2\pi R}{V}$
13	$\dfrac{88}{120}$ or 0.73	50	$\dfrac{4}{18}$	83	$\dfrac{(R_e + h_1)^3}{(R_e + h_2)^3}$
14	$\dfrac{1}{3}$ mi	51	$\dfrac{2 V_0^2}{g} \sin\theta \cos\theta$	84	$2\pi \sqrt{\dfrac{(R_e + h)^3}{g R_e^2}}$
15	zero	52	90°		
16	$\dfrac{2}{60}$	53	constant	85	$\dfrac{4\pi^2}{T^2}(R_e + h)$
		54	1	86	$2\pi(R_e + h)$
17	$\dfrac{20 + 50}{2}$	55	6		
		56	toward center of earth		
18	9.8 m/sec²	57	45°		
19	false	58	$\dfrac{1}{4} g$		
20	20				
21	$6\sqrt{3}$	59	$\dfrac{1}{\sqrt{2}}$		
22	50				
23	88 ft/sec	60	$\dfrac{V_0^2}{g}$		
24	$\dfrac{1}{3} \times \dfrac{88}{60}$				
		61	$\dfrac{1}{4}$		
25	$\dfrac{88}{120}$	62	false		
26	120 sec	63	$z \cos\theta$		
27	N	64	$\dfrac{\frac{1}{2} v^2}{R_e}$		
28	$\dfrac{88}{120}$	65	$2\sqrt{2}$		
29	false	66	$\dfrac{v^2}{r}$		
30	$\dfrac{2}{3} \times \dfrac{88}{60}$				
31	b	67	$\sqrt{8}$		
		68	$2\pi R$		
32	$\dfrac{2}{9}$	69	$\dfrac{1}{2}$		
33	$\dfrac{1}{2} \times \dfrac{88}{60}$	70	increase		
		71	cm		
34	a, b, and c	72	$\dfrac{2\pi R}{v}$		
35	−44				
36	1000				

23

Chapter 3
DYNAMICS

3-1 A car is traveling at 60 mph on a level road. If the sum of all the forces on the car is zero, the car will [continue on at 60 mph ; slow down, but never come to rest ; slow down and come to rest] .   **7**

Newton's first law applies only to bodies at rest [*true* ; *false*] .   **44**

If the net force on a body is zero, then the body must be at rest [*true* ; *false*] .   **20**

If incorrect go to 3-7 ; if correct go to 3-5.

3-2
(from 3-5) An example of the previous situation is a block being pushed against a wall. As long as the *net* force on a body is zero, the body will obey Newton's first law; i.e., remain in a state of rest or constant velocity.

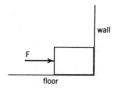

Question: Suppose two forces F_1 and F_2 are acting on M. Then the net force is given by the vector [A ; B ; C ; D] as shown below.   **53**

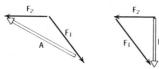

Go to 3-3.

3-3 Suppose three forces F_1, F_2, and F_3 are acting on M.
(from Then the net force is the vector [B ; C ; zero ; 78
3-2)   none of these] .

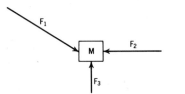

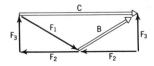

Go to 3-4.

3-4 A tractor is pulling a log at a constant
(from velocity of 2 m/sec with a force of
3-3) $F = 200$ newtons. The downward gravitational force on the log is $F_g = 400$ newtons.

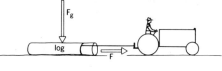

The force of the ground pushing up on the log must then be [0 ; 400 ; 400g] . 70
newtons.

The net force on the log is _____ newtons. 2

The strength of the frictional force of the ground pushing against the direction of
motion of the log is _____ newtons. It is this force that the tractor must overcome in 37
order to be able to pull the log.

The net force on the tractor alone is [0 ; 200 ; neither] newtons. 62

Go to 3-31.

3-5 It can be possible for a body to remain at rest while being pushed by external forces
(from [true ; false] . 45
3-1, 3-7,
or 3-14)

Go to 3-2.

3-6 According to Newton's Second Law the applied force on a body is the product of its mass
(from times its acceleration. [true ; false] . 56
3-31)

If incorrect go to 3-15 ; if correct to to 3-8.

3-7
(from 3-1)

If the net force on a body is zero it can still be moving at constant velocity
[*true* ; *false*] .   16

If a body is not at rest, then the net force acting on it can never be zero
[*true* ; *false*] .   22

If incorrect go to 3-14 ; if correct go to 3-5.

3-8
(from 3-6 or 3-15)

_____ different forces are acting on the block being pushed against a wall by an applied force F_{app}. Do not ignore gravity.   54

The contact force of the floor pushing up is of magnitude
[0 ; Mg ; F_{app} ; $\sqrt{F_{app}^2 + M^2g^2}$; none of these] .   50

The contact force of the wall pushing back on the block is of magnitude
[0 ; Mg ; F_{app} ; $\sqrt{F_{app}^2 + M^2g^2}$; none of these] .   14

Go to 3-9

3-9
(from 3-8)

Consider a collision between M_1 and M_2 where some of the mass of M_2 gets transferred to M_1. Let M_1' and M_2' be the masses after the collision. Then the increase in mass of M_1 is $M_1' - M_1$ which equals

$\left[(M_2 - M_2') \; ; \; (M_2' - M_2) \; ; \; (M_1 - M_2) \; ; \; (M_2' - M_1') \; ; \; \frac{1}{2}(M_2 - M_1) \right]$.   19

According to the law of conservation of momentum,
$M_1 V_1 = M_1' V_1'$ [*true* ; *false*] .   81

If M_2 was initially at rest $M_2' V_2' =$
[$M_1 V_1 + M_1' V_1'$; $M_1(V_1 - V_1')$; none of these] .   76

Go to 3-20.

3-10
(from 3-20)

In the pulley problem to the right, ignore friction and mass of the pulley. The net vertical force on M_1 is [0 ; $M_1 g$; $(M_1 + M_2)g$; $(M_1 - M_2)g$] .
If M_1 is held stationary by hand, the net force on M_1 is _____ .
If M_1 is held stationary, the tension in the string is _____ .
If M_1 is released, the net force of M_1 is [less than ; the same as ; greater than] $M_2 g$.
After M_1 is released the magnitude of its acceleration a_1 is [less than ; greater than ; the same as] a_2.

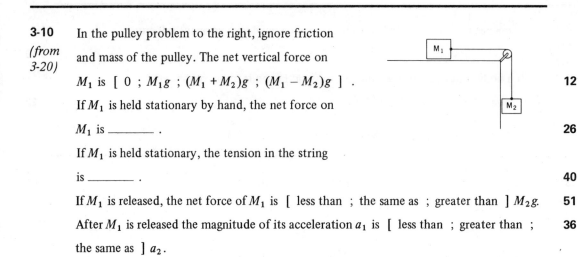

12

26

40

51

36

Go to 3-11.

3-11
(from 3-10)

After M_1 is released the ratio of the net force on M_1 to the net force on M_2 is
[1 ; $\dfrac{g}{a}$; $\dfrac{a}{g}$; $\dfrac{M_1}{M_2}$; $\dfrac{M_2}{M_1}$; none of these] where $a = a_1 = a_2$ is the acceleration.

46

Go to 3-12.

3-12
(from 3-11)

The net force on M_1 is $F_{net} = M_1 a$ and the net force on M_2 is $M_2 a$
[true ; false] .

1

Go to 3-13.

3-13
(from 3-13)

The quantity $(M_1 + M_2)a$ is equal to the net external force acting on the system which is
[$M_1 g$; $M_2 g$; $(M_2 - M_1)g$; none of these] .

The acceleration a is $\left[\dfrac{M_2 - M_1}{M_2 + M_2} g \ ; M_2 g \ ; \dfrac{M_2 g}{M_1 + M_2} \ ; \dfrac{M_2 g}{M_1} \ ; \text{none of these} \right]$.

10

29

Go to 3-18.

3-14 If a body is not at rest, it can be moving with constant velocity and the net force will
(from 3-7) therefore be zero [*true* ; *false*] . 73

If a body is at rest (both v and $a = 0$), then the net force on it must be zero
[*true* ; *false*] . 71

Go to 3-5.

3-15 The key thing to remember about Newton's Second Law is that F in $F = ma$ must always
(from 3-6) be the *net* force. It is not any particular applied force unless that applied force is the only force acting on m.

Go to 3-8.

3-16 Hooke's law states that the restoring force of a spring is proportional to the displacement:
(from 3-24) $F = -kx$. The units of k are [$gm\,cm^2/sec^2$; gm/sec^2 ; $gm\,sec^2/cm$; none of these] . 18

If incorrect go to 3-29 ; if correct go to 3-26.

3-17 The units of $\frac{M}{k}$ are _____ . 67
(from 3-27)

Go to 3-32.

3-18 A mass M is sliding down a frictionless table tilted at
(from 3-13) an angle θ from the horizontal. The contact force of
the table pushing the mass is pointing [left along the
table ; straight up ; up, but perpendicular to the table] .
In the force diagram for this problem the contact force
of the table on the block is [$\vec{F}_1$; $\vec{F}_2$; $-\vec{F}_2$; $\vec{F}_3$;
$-\vec{F}_3$] .
The force of the block pushing against the table is [$\vec{F}_2$; $-\vec{F}_3$;
neither] .
The previous question is an application of Newton's [First ;
Second ; Third] Law of motion.
The magnitude of F_2 is Mg [true ; false] .

In this right triangle $\sin\theta = \dfrac{y}{z}$, $\cos\theta = \dfrac{x}{z}$, and $\tan\theta = \dfrac{y}{x}$ by definition.

Hence $F_1 = \left[\; F_2\cos\theta \;;\; F_2\sin\theta \;;\; \dfrac{F_2}{\cos\theta} \;;\; \dfrac{F_2}{\sin\theta} \;;\; \text{none of these}\;\right]$.

The net force on the block is [F_1 ; F_2 ; F_3 ; none of these] .
The net force on the block is [$Mg\sin\theta$; $Mg\cos\theta$; neither] .
The net force on the block has the value [Ma ; $Ma + Mg\cos\theta$;
$\sqrt{(Ma)^2 + (Mg\cos\theta)^2}$; none of these] . The answer can be equated to the previous
answer [true ; false] .
The acceleration is $a =$ _____ .

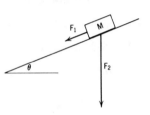

80
49
75
32
21
25
74
42
9
28
58

Go to 3-19.

3-19 Consider a pendulum bob of mass M swinging around in
(from 3-18) a circle of radius R.

The net force on M is pointed [down; along the string ;
toward the center of the circle] .

The net force has the value $\dfrac{MV^2}{R}$ [true ; false] .

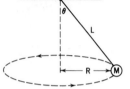

69
48

Go to 3-21.

3-20
(from 3-9)

The law of conservation of momentum states that
$M_1V_1 + M_2V_2 = M_1'V_1' + M_2'V_2'$ [true ; false] . 57

Go to 3-10.

3-21
(from 3-19)

The net force has the value $\left[Mg \sin\theta \ ; \ \dfrac{Mg}{\sin\theta} \ ; \ Mg \tan\theta \ ; \ Mg \cos\theta \ ; \ \text{none of these} \right]$. 3

Go to 3-25.

3-22
(from 3-21 or 3-19)

We wish to obtain the period of oscillation of this so-called circular pendulum.

The centripetal acceleration of M has the value $\dfrac{4\pi^2 R}{T^2}$ [true ; false] . 52

The centripetal acceleration also has the value $g \tan\theta$ [true ; false] . 79

These two quantities can be equated. Solving this equation for T gives $T = 2\pi\sqrt{\dfrac{R}{g \tan\theta}}$

Note that $\dfrac{R}{L} = \sin\theta$, or $R = L \tan\theta \cos\theta$; hence $T = \left[2\pi\sqrt{\dfrac{L}{g}} \ ; \ 2\pi\sqrt{\dfrac{L \cos\theta}{g}} \ ; \ 2\pi\sqrt{\dfrac{L}{g \cos\theta}} \ ; \ \text{none of these} \right]$. 17

If θ is kept small, $\cos\theta$ is close to one and we then have the good approximation that
$T = 2\pi\sqrt{\dfrac{L}{g}}$ [true ; false] . 31

Now look at the motion in the horizontal direction only (look at V_x), such a motion is mathematically identical to a simple pendulum. Hence the above is a derivation of the formula $T = 2\pi\sqrt{\dfrac{L}{g}}$ for a simple pendulum of small oscillation.

Go to 3-23.

3-23
(from 3-22)

We have the formula for the acceleration of a pendulum bob (or any other form of SHM) in terms of its displacement and period. The formula is $a = \dfrac{4\pi^2}{T^2} x$.

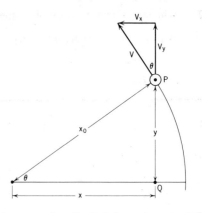

Now we would like to obtain a formula for the velocity in terms of x and T. Let us refer to the reference circle where point P is in circular motion and point Q is in SHM. The velocity of point Q will then be [V ; V_x ; V_y ; none of these] . **60**

By similar triangles $\dfrac{V_x}{V} =$ _____ . Hence $V_x = V \dfrac{\sqrt{x_0^2 - x^2}}{x_0}$ **55**

In the reference circle V is the circumference divided by the period [*true* ; *false*] . **77**

The maximum velocity $V =$ $\left[\dfrac{x_0}{T} \; ; \; \dfrac{2\pi x_0}{T} \; ; \; \dfrac{\pi x_0}{T} \; ; \; \text{none of these} \right]$. **6**

Hence $V_x =$ $\left[\dfrac{2\pi}{T} \sqrt{1 - \left(\dfrac{x}{x_0}\right)^2} \; ; \; \dfrac{2\pi}{T} \sqrt{x_0^2 - x^2} \; ; \; \dfrac{2\pi}{T} \dfrac{\sqrt{x_0^2 - x^2}}{x_0^2} \; ; \; \text{none of these} \right]$. **39**

Go to 3-28.

3-24
(from 3-30)

In the previous frame we had $\dfrac{4\pi^2}{T^2} = 0.66 \times 10^{30}$. Solve this for T. The result is

$T =$ _____ sec. **11**

This is a frequency of _____ oscillations per sec. A charged particle oscillating at this frequency will emit electromagnetic radiation of this frequency. Such radiation appears as infrared radiation. Hence heated hydrogen gas emits infrared radiation at this precise frequency. If the frequency had been $3\tfrac{1}{2}$ times greater, it would appear as visible light. **65**

Go to 3-16.

3-25 The forces and force triangle for the pendulum bob are shown to
(from   the right.
3-21)

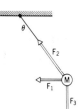

F_2 is the [gravitational force ; net force ; tension in the string] . 33

F_1 is the [net force ; gravitational force ; neither] . 4

$F_3 = Mg$ [true ; false] . 61

The ratio $\dfrac{F_1}{F_3}$ = [$\sin\theta$; $\cos\theta$; $\tan\theta$] . 43

The ratio $\dfrac{F_{net}}{Mg}$ = [$\sin\theta$; $\cos\theta$; $\tan\theta$] . 34

This ratio is also the same as $\left[\dfrac{a}{g}\ ;\ \dfrac{g}{a}\ ;\ \dfrac{a\sin\theta}{g}\ ;\ \text{none of these}\right]$ 38

The acceleration of the bob M has the magnitude $a = g\tan\theta$

[true ; false] . 13

Depending on θ, the centripetal acceleration could be greater than g

[true ; false] . 72

Go to 3-22.

3-26 The period of oscillation of a mass M pulled by a spring is related to the force constant k
(from   by the equation $\dfrac{4\pi^2}{T^2} = \left[k\ ;\ \dfrac{k}{M}\ ;\ \dfrac{M}{k}\ ;\ 2\pi\sqrt{\dfrac{M}{k}}\ ;\ \sqrt{\dfrac{M}{k}}\ ;\ \text{none of these}\right]$. 47
3-16 or
3-29)

Go to 3-27.

3-27 The acceleration of M is $a = \left[-kx\ ;\ -\dfrac{kx}{M}\ ;\ \text{neither}\right]$. 23
(from
3-26)   The acceleration of M also has the value $a = -\dfrac{4\pi^2}{T^2}x$ [true ; false] . 30

Hence $\dfrac{4\pi^2}{T^2} = \dfrac{k}{M}$ [true ; false] . 8

Go to 3-17

3-28
(from 3-23)

In Curve A the time [t_1 ; t_2 ; t_3 ; none of these] corresponds to one complete period.

Suppose the displacement of an object in SHM plotted as a function of time looks like Curve A. Then the acceleration plotted as a function of time will look like Curve _____ .

The net force will look like Curve _____ .

The velocity is represented by Curve _____ .

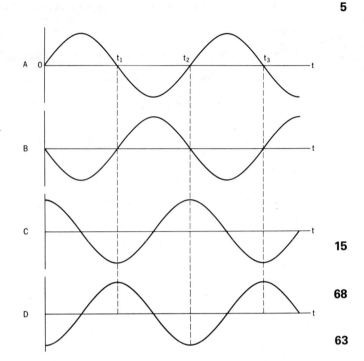

5

15

68

63

Go to 3-30.

3-29
(from 3-16)

The force constant $k = \dfrac{F}{x}$ and has units of dynes/cm. But dynes (mass times acceleration) has units _____ .

35

Go to 3-26.

3-30
(from 3-28)

If two hydrogen atoms are pushed together, there will be a repulsive force between the two protons looking something like Fig. 1. On the other hand, the attractions between the electrons and the two protons will contribute to an attractive force looking something like Fig. 2. The sum of the two curves gives a net force between the two hydrogen atoms something like Fig. 3. The equilibrium position has the value $r_0 = 4 \times 10^{-9}$ cm. At this position the slope of the curve is $\dfrac{\Delta F}{\Delta r} = -1.1 \times 10^6$ dynes/cm. If the two protons are pushed 10^{-9} cm closer together (now $r = 3 \times 10^{-9}$ cm), the net force on each of them will be [zero ; repulsive ; attractive] .

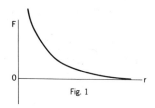
Fig. 1

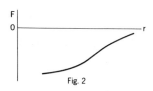
Fig. 2

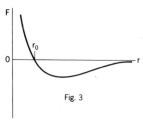
Fig. 3

r is distance of proton from center of molecule

The value of this force will be _____ dynes.

Let f be the value of this restoring force and x the displacement from the equilibrium position. Then $f = -(1.1 \times 10^6)x$ in dynes [true ; false] .

If the mass of a proton is 1.67×10^{-24} gm, the relation between acceleration and displacement will be $a = -(0.66 \times 10^{30})x$ in cm/sec^2 [true ; false] .

If T is the period of oscillation of the two protons in a hydrogen molecule, then

$$a = -\frac{4\pi^2}{T^2} x \quad [\text{ true ; false }] .$$

The quantity $\dfrac{4\pi^2}{T^2}$ has the value _____ .

66

27

41

24

59

64

Go to 3-24

3-31
(from 3-4)

Review Questions: We see that Newton's First Law applies to bodies moving with constant velocity as well as at rest. Also Newton's First Law only applies if the vector sum of all the external forces is zero. Now for a short quiz on this.

1. If a body is at rest, each external force acting on it must necessarily be zero. True or false? _____

2. The net force on a body will always be zero if the body is _____.

3. In order for any body to move with constant velocity, there must not be any frictional forces. True or false? _____

4. Check the correct choices:

Newton's First Law holds true for

_____ stationary observers.

_____ observers moving at constant velocity.

_____ accelerating observers.

_____ all possible observers.

A body will move with constant velocity if

_____ there is a constant net force pushing it.

_____ the vector sum of all the external forces is zero.

_____ none of the above.

Go to 3-6.

3-32 Review Questions:

(from 3-17)

1. The net force on any mass M must always be M times the acceleration of M.

   true _____

   false _____

2. Ignoring rotation of the earth, the net force on a mass M sitting on a table is _____ .

3. The net force on a low-flying earth satellite of mass M is _____ .

4. List the different forces acting on the second car of a train.

   1. _____ , 2. _____ , 3. _____ ,
   4. _____ , 5. _____

5. In SHM what is the ratio of the displacement to the acceleration? Give answer in terms of the period T. (Hint: see frame 3-27).

   $$\frac{x}{a} = \underline{\qquad\qquad} .$$

6. If $F = \frac{1}{C} x$, what are the units of C in gm, cm, and sec?

   Units of C = _____ .

Chapter 3 Answer Sheet

Item Number	Answer
1	true
2	zero
3	$Mg \tan \theta$
4	net force
5	t_2
6	$\dfrac{2\pi x_0}{T}$
7	continue on at 60 mph
8	true
9	Ma
10	$M_2 g$
11	7.74×10^{-15}
12	zero
13	true
14	F_{app}
15	B
16	true
17	$2\pi \sqrt{\dfrac{L \cos\theta}{g}}$
18	gm/sec²
19	$M_2 - M_2'$
20	false
21	true
22	false
23	$-\dfrac{kx}{M}$
24	true
25	$F_2 \sin \theta$
26	zero
27	1.1×10^{-3}
28	true
29	$\dfrac{M_2 g}{M_1 + M_2}$
30	true
31	true
32	Third
33	tension
34	$\tan \theta$
35	gm cm/sec²
36	the same as
37	200
38	$\dfrac{a}{g}$
39	$\dfrac{2\pi}{T}\sqrt{x_0^2 - x^2}$
40	$M_2 g$
41	true
42	$Mg \sin \theta$
43	$\tan \theta$
44	false
45	true
46	$\dfrac{M_1}{M_2}$
47	$\dfrac{k}{m}$
48	true
49	$\vec{F}_3$
50	Mg
51	less than
52	true
53	B
54	4
55	$\dfrac{y}{x_0}$
56	false
57	true
58	$g \sin \theta$
59	true
60	V_x
61	true
62	zero
63	C
64	$.66 \times 10^{30}$
65	1.29×10^{14}
66	repulsive
67	sec²
68	B
69	toward the center
70	400
71	true
72	true
73	true
74	F_1
75	$-\vec{F}_3$
76	none
77	true
78	zero
79	true
80	up, but perpendicular to the table
81	false

Chapter 4
GRAVITATION

4-1 Suppose Newton had defined a physical quantity Q called gravitational charge which is to be proportional to the gravitational force and to be the source of gravity. His universal law of gravitation would then be $F = \dfrac{Q_1 Q_2}{r^2}$ where

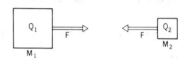

Q_1 is the amount of gravitational charge associated with mass M_1 and Q_2 is the amount associated with M_2. Using CGS units, consider two identical bodies each with one unit of Q separated by 1 cm. According to the equation $F = \dfrac{Q^2}{r^2}$ the force of attraction would be _____ dynes. **30**

If incorrect go to 4-7 ; if correct go to 4-2.

4-2
(from 4-1 or 4-8)
In order to give a force of 1 dyne at a separation of 1 cm the masses of these two identical bodies would have to be $\left[1 \ ; \ G \ ; \ \dfrac{1}{G} \ ; \ \sqrt{G} \ ; \ \dfrac{1}{\sqrt{G}} \ ; \ \text{none of these} \right]$. (Use the relation $F = \dfrac{GM^2}{r^2}$.) **48**

Go to 4-8.

4-3
(from 4-8)
The fixed relation between gravitational charge (the property of a body responsible for gravitational force) and mass (the ratio $\frac{F_{net}}{a}$ whenever a body is accelerated) would then be $Q = \left[M \ ; \ \sqrt{G}M \ ; \ \frac{1}{\sqrt{G}}M \ ; \ GM \ ; \ \frac{M}{G} \right]$. We would obtain the experimental result that gravitational charge and inertial mass are always strictly proportional to each other. Such a result is a statement of the principle of equivalence [*true* ; *false*].

39

12

Go to 4-4.

4-4
(from 4-3)
If Newton had instead defined gravitational charge by the equation $F = G\frac{Q_1 Q_2}{r^2}$ where G has the value $G = 6.67 \times 10^{-8}$ in CGS units, then the relation between Q and M would be $Q = $ _____ .

18

Go to 4-5

4-5
(from 4-4)
This is in effect what Newton did, and the result $Q = M$ is what we call the principle of equivalence. On the basis of experiment we can replace the quantity called gravitational charge (also called gravitational mass) with the quantity mass (also called inertial mass). The main point to be made here is that until the experiments were performed, it should not have been obvious that the two would always be the same. The experimental fact that the two do turn out to be the same should be interpreted as a most fundamental law of nature of profound significance.

Go to 4-9

4-6
(from 4-21 or 4-31)
The centripetal force has the same value as $\frac{GM_e m}{R_1^2}$ [*true* ; *false*].

56

Go to 4-20.

42

4-7
(from 4-1)

If we substitute $Q = 1$ into the equation $F = \dfrac{Q^2}{r^2}$ we obtain $F = \left[\ \dfrac{1}{r^2}\ ;\ \dfrac{G}{r^2}\ ;\ \text{neither}\ \right]$.   62

If we also substitute $r = 1$ into the same equation we obtain $F = \dfrac{1}{1^2} = 1$ dyne.

Go to 4-2.

4-8
(from 4-2)

In the equation $F = \dfrac{GM^2}{r^2}$, if both F and r are 1, we obtain $1 = \dfrac{GM^2}{1}$ [*true* ; *false*].   5

or $M^2 = $ _____ .   32

Go to 4-3.

4-9
(from 4-5)

If the mass of the earth were doubled and its radius kept the same, everybody's
[weight ; mass] would [double ; triple ; quadruple].   37
50

If the mass of the earth were kept the same, but the radius halved, everybody's weight
would [halve ; remain the same ; double ; none of these].   21

If incorrect go to 4-17 ; if correct go to 4-12.

4-10
(from 4-27)

If the radius of its orbit is 4 times smaller, the circumference will be _____ times   51
smaller. If the circumference is this much smaller and the speed is twice as fast, the
moon would travel the circumference _____ times as fast. Hint: $v = 2\pi\dfrac{R}{T}$ or $T = 2\pi\dfrac{R}{v}$.   9

Go to 4-28.

4-11
(from 4-34)

The weight of a man in free fall is [0 ; Mg ; Ma ; neither].   16

Go to 4-42.

4-12
(from 4-9 or 4-17)
An earth satellite of mass m is in a circular orbit of radius R_1 and velocity v_1.
The satellite is accelerating
[*true* ; *false*].

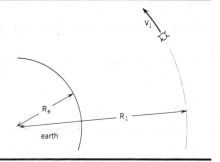

28

If incorrect go to 4-18 ; if correct go to 4-13.

4-13
(from 4-12)
The direction of the satellite's acceleration is
[A ; B ; C ; D ; none of these] .

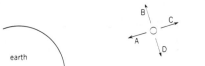

41

Go to 4-19.

4-14
(from 4-19)
The gravitational force acting on the satellite is $\left[\dfrac{M_e M}{R_1^2} \; ; \; \dfrac{GM_e m}{R_e^2} \; ; \; \dfrac{GM_e m}{R_1} \; ; \; \text{none of these} \right]$ where M_e is the mass of the earth.

59

If incorrect go to 4-22 ; if correct go to 4-15.

4-15
(from 4-14 or 4-22)
Both the gravitational force and the centripetal force act independently on m; i.e., the net force is the sum of the two [*true* ; *false*].

22

If incorrect go to 4-23 ; if correct go to 4-21.

4-16
(from 4-20)
Another earth satellite has a circular velocity v_2 at a distance R_2. Then $\dfrac{v_2}{v_1} = $ _____ . Give answer in terms of R_1 and R_2.

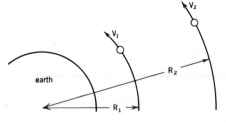

8

Go to 4-26.

4-17
(from 4-9)
If $F_g = G\dfrac{M_e M}{R_e^2}$; and $F_g' = G\dfrac{M_e M}{\left(\frac{1}{2}R_e\right)^2}$, then $\dfrac{F_g'}{F_g} = $ _____ .

1

Go to 4-12.

4-18
(from 4-12)
Any object moving uniformly in a circle must have an acceleration $a_c = \dfrac{v^2}{r}$. You apparently still do not understand centripetal acceleration. If this is the case, we recommend that you go back and rework the Chapter 2 and Chapter 3 programs without looking at your previous work.

Go to frame 2-1 of Chapter 2.

4-19
(from 4-13)
The centripetal force acting on the satellite is $\left[\dfrac{v_1^2}{R_1}\ ;\ \dfrac{mv_1^2}{R_1}\ ;\ \dfrac{mv_1^2}{R_1 - R_e}\ ;\ \dfrac{mv_1^2}{R_e}\ ;\right.$ none of these $\Big]$.   33

Go to 4-14.

4-20
(from 4-6)
$\dfrac{v_1^2}{R_1} = \dfrac{GM_e}{R_1^2}$ [*true* ; *false*].   47

Then $v_1 = \left[\dfrac{GM_e}{R_1}\ ;\ \sqrt{\dfrac{GM_e}{R_1^3}}\ ;\ \sqrt{\dfrac{GM_e}{R_1}}\ ;\ \dfrac{GM_e}{R_1^3}\ ;\ \text{none of these}\right]$.   15

Go to 4-16.

4-21
(from 4-15 or 4-24)
The centripetal force required to keep the satellite on orbit is supplied by [rocket thrust ; gravitational attraction of the earth ; neither, it is zero].   60

Go to 4-6.

4-22
(from 4-14)
The gravitational force is obtained by substituting $R = R_1$ into the formula

$F = \dfrac{GM_e m}{R^2}$ [*true* ; *false*].   55

Go to 4-15.

4-23
(from 4-15)

You are having serious trouble in understanding the basic principle of earth satellites. Also your answer violates the equation $F_{net} = ma$. You must have claimed either that

$$F_{net} = \frac{GM_e m}{R_1^2} + \frac{mv_1^2}{R_1} \text{ or that } F_{net} = 0.$$

But Newton claims that $F_{net} = ma$ where $a = \dfrac{v_1^2}{R_1}$ is the acceleration. Of course the question arises: Where does F_{net} come from? The answer is gravity; and it has the value $G\dfrac{M_e m}{R_1^2}$.

Go to 4-31.

4-24
(from 4-31)

The discussion in the frame above (4-23) just got through telling you that F_{net} is supplied by gravity. This program recommends that you start over with Chapter 2. If you feel you understand about the forces on an earth satellite you may go on to frame 4-21.

Go to Chapter 2 or to 4-21.

4-25
(from 4-38)

If $v^2 = 4gR$, then the acceleration $\dfrac{v^2}{R} = 4g$ [*true* ; *false*]   46

Go to 4-39.

4-26
(from 4-16)

If the moon's orbit were moved to such a radius that its new circular velocity would be twice its present velocity, the radius of the new orbit would be $\left[\dfrac{1}{4} \; ; \; \dfrac{1}{2} \; ; \; 2 \; ; \; 4 \; ; \; \text{none of these}\right]$ times the present radius.   3

Go to 4-27.

4-27
(from 4-26)

The period of this new orbit would be _____ times its present period.   25

If incorrect go to 4-10 ; if correct go to 4-28.

4-28
(from 4-27 or 4-10)
Kepler's third law states $\left(\frac{T'}{T}\right)^2 = \left(\frac{R'}{R}\right)^3$. In the preceding problem the ratio of the distances is $\frac{R'}{R} = \frac{1}{4}$. Then the quantity $\left(\frac{T'}{T}\right)^2$ will be _____ . **19**

The quantity $\left(\frac{R'}{R}\right)^{\frac{3}{2}}$ is $\frac{T'}{T}$ [*true* ; *false*]. **58**

The quantity $\left(\frac{R'}{R}\right)^{\frac{3}{2}}$ has the value _____ . **38**

Go to 4-29.

4-29
(from 4-28)
If the moon had twice its present mass, and stayed in a circular orbit at its present distance, the period of this heavier moon would be _____ times its present period. **10**

If incorrect go to 4-37 ; if correct go to 4-38.

4-30
(from 4-33 or 4-43)
The weight of a man is always proportional to his acceleration [*true* ; *false*]. **17**

Go to 4-34.

4-31
(from 4-23)
The centripetal force required to keep the satellite on orbit is supplied by
[rocket thrust ; gravitational attraction of the earth ; neither, it is zero] **43**

If incorrect go to 4-24 ; if correct go to 4-6.

4-32
(from 4-45)
[Kepler ; Newton ; Galileo] used logical [induction ; deduction] to generalize from observations on Mars that all planets must move in [ellipses ; circles] . **61**
7
20

Go to 4-33.

4-33
(from 4-32)
The force of gravity on a body is always proportional to its acceleration
[*true* ; *false*]. **14**

If incorrect go to 4-43 ; if correct go to 4-30.

4-34
(from 4-30)
The weight of a man is always proportional to the force of gravity acting on him [*true* ; *false*]. In this manual we use a physiological definition of weight; e.g., a passenger in a non-rotating earth satellite has zero weight.

26

If incorrect go to 4-11 ; if correct go to 4-42.

4-35
(from 4-40)
If the net force on an earth satellite were zero, it would travel in a straight line rather than a circle [*true* ; *false*].

53

Go to 4-36.

4-36
(from 4-35)
The net force on an earth satellite is the vector sum of the gravitational and the centripetal force [*true* ; *false*].

57

Go to 4-41.

4-37
(from 4-29)
Kepler's third law says the period depends on the mass of the satellite [*true* ; *false*].

2

A man-made satellite in a circular orbit of $R = 240,000$ mi (same as that of the moon) would have the same period as that of the moon [*true* ; *false*].

35

Go to 4-38.

4-38
(from 4-29 or 4-37)
A roller-coaster car has a peak velocity of $v = \sqrt{4gR}$ where R is the radius of curvature of the dip. The acceleration of this when at the bottom of the dip is _____ .

13

If incorrect go to 4-25 ; if correct go to 4-39.

4-39
(from 4-38 or 4-25)
Consider a man of mass M sitting in the car. The net force on this man is _____ .

54

If incorrect go to 4-44 ; if correct go to 4-45.

4-40
(from 4-42)
Since the net force on an earth satellite is the vector sum of the gravitational force and the centripetal force, the net force is zero [*true* ; *false*].   40

If incorrect go to 4-35 ; if correct go to 4-46.

4-41
(from 4-36)
Consider an earth satellite in circular orbit.
The net force is equal to the gravitational force [*true* ; *false*].   23
The net force is equal to the centripetal force [*true* ; *false*].   4

Go to 4-46.

4-42
(from 4-34 or 4-11)
Could Newton have measured the mass of Jupiter? Newton knew from Galileo's work that Io, a moon of Jupiter, had a period of 1.77 days or $T = 1.53 \times 10^5$ sec. The radius of Io's orbit is $R = 4.22 \times 10^5$ km. If M_j is the mass of Jupiter, then the centripetal acceleration of Io is $\left[\dfrac{4\pi^2}{T^2}R^2 \; ; \; \dfrac{4\pi^2}{T^2 R} \; ; \; \dfrac{4\pi^2 R}{T^2} \; ; \; \text{none of these} \right]$.   31

This centripetal acceleration must equal the acceleration due to gravity from Jupiter.
[*true* ; *false*].   34

The acceleration due to Jupiter's gravity at the position of Io is $\left[\dfrac{GM_j M_{Io}}{R^2} \; ; \; \dfrac{GM_j}{R} \; ; \; \right.$ 6
$\left. \dfrac{GM_j}{R^2} \; ; \; \text{none of these} \right]$. Equate the two previous results to obtain the mass of Jupiter.
$M_j =$ _____ .   42
The numerical result is $M_j =$ _____ kg.   11

Go to 4-40.

4-43
(from 4-33)
The gravitational force on a mass M at rest on the surface of the earth is
$Fg = \left[0 \; ; \; Mg \; ; \; GM_e M \; ; \; \dfrac{GM_e}{R_e^2} \; ; \; \text{none of these} \right]$.   24

Go to 4-30.

4-44
(from 4-39)
Remember $F_{net} = Ma$. If $a = 4g$, then F_{net} must be M times $4g$ [*true* ; *false*]. By now you have probably realized that $F = Ma$ is not as simple as it sounds.   52

Go to 4-45.

4-45
(from 4-39 or 4-44)

Lef F' be the force of the seat pushing up on the man. Then $F' = Ma = 4Mg$
[*true* ; *false*].   27

There is a force of gravity pulling down on the man of magnitude $F_g = Mg$
[*true* ; *false*].   29

The magnitude of the net force on the man is [$F' + F_g$; $F' - F_g$; neither] .   45

$F' - F_g$ = [2 ; 3 ; 4 ; 5] Mg.   36

The force of the seat pushing up on the man is [3 ; 4 ; 5] Mg.   44

The man will feel [2 ; 3 ; 4 ; 5 ; none of these] times his normal weight.   49

Go to 4-32.

4-46
(from 4-40 or 4-41)

Review Questions:

This is a short quiz on Chapter 4. Use pencil.

1. According to the principle of equivalence, gravitational charge should be exactly proportional to _____ .

2. What quantities need to be known in order to calculate the mass of the sun? (see frame 4-42) _____ .

3. Weight is always equal to gravitational force.   true _____
                                                                            false _____

4. The period of a satellite (increases ; decreases) with increasing radius of the satellite orbit. The period (increases ; decreases ; remains the same) with increasing mass of the satellite. (Underline correct choice.)

5. For an earth satellite in circular orbit the net force is $\dfrac{mv^2}{R}$.   true _____
                                                                                                     false _____

Chapter 4 Answer Sheet

Item Number	Answer
1	4
2	false
3	$\frac{1}{4}$
4	true
5	true
6	$\frac{GM_j}{R_2}$
7	induction
8	$\sqrt{\frac{R_1}{R_2}}$
9	8
10	1
11	1.9×10^{27}
12	true
13	$4g$
14	false
15	$\sqrt{\frac{GM_e}{R_1}}$
16	zero
17	false
18	M
19	$\frac{1}{64}$
20	ellipses
21	none
22	false
23	true
24	Mg
25	$\frac{1}{8}$
26	false
27	false
28	true
29	true
30	1
31	$\frac{4\pi^2 R}{T^2}$

Item Number	Answer
32	$\frac{1}{G}$
33	$\frac{mv_1^2}{R_1}$
34	true
35	true
36	4
37	weight
38	$\frac{1}{8}$
39	$\sqrt{GM}$
40	false
41	A
42	$\frac{4\pi^2 R^3}{GT^2}$
43	gravitational attraction
44	5
45	$F' - F_g$
46	true
47	true
48	$\frac{1}{\sqrt{G}}$
49	5
50	double
51	4
52	true
53	true
54	$4Mg$
55	true
56	true
57	false
58	true
59	none
60	gravitational attraction
61	Kepler
62	$\frac{1}{r^2}$

Chapter 5
ANGULAR MOMENTUM AND ENERGY

5-1 The law of conservation of momentum states that the vector sum of the momenta of any two objects must always remain the same [*true* ; *false*] .   **130**

Go to 5-6.

5-2
(from 5-6)
The angular momentum of mass M with respect to O is [$MR_1 V_\perp$; $MR_2 V$; $MR_2 V_\perp$] .   **42**

The angular momentum is also equal to $R_1 MV$ [*true* ; *false*] .   **23**

If incorrect go to 5-7 ; if correct go to 5-3.

5-3
(from 5-2 or 5-7)
A man holds a spinning bicycle wheel above his head (the axis of the wheel is vertical). As viewed from above the wheel is rotating clockwise. The man now steps onto a frictionless turntable (a platform free to revolve).

Ignoring any friction of air resistance, the turntable will start to rotate [clockwise ; counterclockwise ; neither] as seen from above.   **78**

If the man stops the bicycle wheel by grasping the rim he will start to rotate [clockwise ; counterclockwise ; neither] .   **95**

If incorrect go to 5-9 ; if correct go to 5-4.

5-4 The man steps onto the turntable with a bicycle wheel at rest (not rotating). While on the
(from turntable he starts the wheel spinning clockwise as viewed from above by pushing the rim
5-3 or with his hand.
5-9)

He will then spin [clockwise ; counterclockwise ; neither] .   **6**

Except for direction, the angular momentum of the man will be [the same as ; more   **108**
than ; less than] that of the wheel.

Next the man turns the wheel upside down so that it is rotating counterclockwise with
the same speed as before. The man will now be rotating [clockwise ; counterclock-   **52**
wise ; neither] with the [same ; twice ; neither] speed of rotation as before.   **91**

Finally the man grasps the rim of this upside-down wheel and stops its rotation. Now in
which direction will the man rotate? _____ . He will rotate with the same   **33**
speed of rotation as when he was initially holding the spinning wheel over his head.
[*true* ; *false*]   **20**

If any of the responses in this frame are incorrect go to 5-5 ; if all are correct go to 5-37.

5-5 In the previous problem when the man first stepped on the turntable with a nonrotating
(from wheel, the angular momentum of the system (wheel plus man) was zero [*true* ; *false*] .   **82**
5-4) At all subsequent stages in the problem the total angular momentum of the system was
zero [*true* ; *false*] .   **100**

Go to 5-8.

5-6 By a closed system we mean that the particles must not be under the influence of
(from [internal ; external] forces.   **142**
5-1) The vector sum of the momenta of any two objects will remain the same if they are not
under the influence of any external forces [*true* ; *false*] .   **75**

Go to 5-2.

5-7
(from 5-2)

Triangle OBM is [congruent ; similar ; neither] to the small right triangle which has V as a hypotenuse.

Hence $\dfrac{V_\perp}{V} = $ _____ .

Then $R_1 V = R_2 V_\perp$

And $MR_1 V = MR_2 V_\perp$ which is the angular momentum of M [true ; false] .

90

17

11

Go to 5-3.

5-8
(from 5-5)

In the winter the earth is closer to the sun. At that time of the year its orbital velocity is [greater ; less ; the same] as it is in the summer.

In winter the angular momentum of the earth is [greater ; less ; the same] as it is in summer.

132

53

Go to 5-37.

5-9
(from 5-3)

The initial angular momentum of the man plus the wheel was [clockwise; counterclockwise ; zero] .

After grasping the rim, the angular momentum of the man plus wheel will [increase; decrease ; remain the same].

80

68

Go to 5-4.

5-10
(from 5-37 or 5-48)

The dyne has the units of _____ .

$\left(\dfrac{\text{Angular momentum}}{\text{sec}} \right)$ has the same units in gm, cm and sec as torque [true ; false] .

Torque has the same units in gm, cm, and sec as energy [true ; false] .

Dynes/cm has the same units as ergs [true ; false] .

Dynes times cm has the same units as ergs [true ; false] .

79

47

12

1

57

Go to 5-11.

5-11
(from 5-10)
If work is done on a mass M, the kinetic energy of M will necessarily increase [true ; false].

84

If incorrect go to 5-22 ; if correct go to 5-12.

5-12
(from 5-11 or 5-22)
In the absence of friction the sum of kinetic plus potential energy will always remain constant [true ; false].

61

If incorrect go to 5-23 ; if correct go to 5-24.

5-13
(from 5-24 or 5-17)
Consider a block of mass M sliding along a frictionless table with velocity V_o. It then collides with an identical mass hanging from a string. Assume an elastic collision (no net loss in KE during the collision). After the collision the KE of the mass on the string is _____ . This mass swings to a maximum height h.

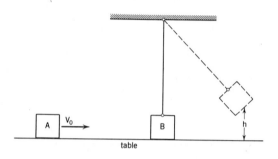

128

Then its potential energy is [Mgh ; $\frac{1}{2}Mgh$; neither].

92

At this moment its potential energy also has the value $\left[\frac{1}{2}MV_o^2 \ ; \ \frac{1}{4}MV_o^2 \ ; \ MV_o \ ; \right.$

140

$\left. \text{none of these} \right]$. The height h has the value $h =$ _____ .

119

Go to 5-14.

5-14
(from 5-13)

Now repeat the above problem assuming a completely inelastic collision (the two masses stick together after the collision).

The total KE just after the collision will be [less than ; greater than ; the same as] the total KE just before the collision. The total momentum just after the collision will be [greater than ; less than ; the same as] the total momentum just before the collision.

The total momentum after the collision will be _____ .

The velocity of block A just after the collision will be _____ .

The KE of block A just after the collision will be _____ times its original KE.

43

99

121

103

126

If incorrect go to 5-25 ; if correct go to 5-15.

5-15
(from 5-14 or 5-25)

The total loss in KE during the collision is _____ . Mechanical energy is [lost ; increased ; conserved] during this inelastic collision. If h' is the maximum height reached by the two blocks after this inelastic collision, the total potential energy at this position is _____ .

This amount of potential energy has the value $\left[\dfrac{MV_o^2}{8} \; ; \; \dfrac{1}{4}MV_o^2 \; ; \; \dfrac{1}{2}MV_o^2\right]$. It also has the value $2Mgh'$ [*true* ; *false*] .

Hence the height $h' =$ _____ .

The amount of heat energy produced in this collision is $\left[0 \; ; \; \dfrac{MV_o^2}{8} \; ; \; \dfrac{1}{4}MV_o^2 \; ; \; \dfrac{1}{2}MV_o^2 \; ; \; \text{none of these}\right]$. This amount of heat also has the value $\left[\left(\dfrac{1}{2}MV_o^2 - Mgh'\right) \; ; \; 0 \; ; \; \left(\dfrac{1}{2}MV_o^2 - 2Mgh'\right) \; ; \; \text{none of these}\right]$.

At the maximum height h', the potential energy plus [heat energy ; kinetic energy] equals the initial kinetic energy.

2

64

38

104

14

71

26

110

122

Go to 5-27.

5-16
(from 5-28)
The problem you just missed is exactly the same as the roller coaster problem of Chapter 4 which you have already worked. (See Chapter 4 frame 4-38.)

There are _____ external forces acting on the man.   87

The force of gravity is pulling down on him with strength [0 ; 1 ; 2 ; 4] mg.   9

The force of the floor or seat is pushing up with such a strength that the result is

F_{net} = _____ mg.   51

The force of the seat is F_{seat} = _____ mg.   116

The apparent weight of the man is the force with which his body pushes on the seat
[*true* ; *false*] .   137

The apparent weight of the man has the same value as F_{net} [*true* ; *false*] .   48

The apparent weight of the man has the same value as F_{seat} [*true* ; *false*] .   94

Go to 5-30.

5-17
(from 5-24)
What about friction? Kinetic energy can be converted into heat energy [*true* ; *false*] .   124

Potential energy can be converted into heat energy [*true* ; *false*] .   136

Go to 5-13.

5-18
(from 5-31)
The gravitational potential energy at a distance r from the earth's surface has the value $U = mgR^2 \left(\dfrac{1}{R} - \dfrac{1}{r} \right)$ where R is the radius of the earth.

Curve _____ is a plot of this potential energy.   102

The maximum possible value of U is _____ .   62

U_{max} also has the value _____ where V_o is the velocity of escape.   40

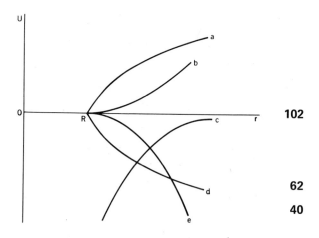

Go to 5-19.

5-19 The energy necessary to put a nosecone into orbit is _____ times the energy necessary 13
(from 5-18) to give escape velocity to the same nosecone.

If incorrect go to 5-32 ; if correct go to 5-21.

5-20 A given amount of energy in any form will always have the same mass value
(from 5-29) [*true* ; *false*] . 115

Go to 5-33.

5-21 One finds in some physics books the statement "Mass can be converted into energy."
(from 5-19 or 5-32) Strictly speaking, this statement is incorrect [*true* ; *false*] . 117

Go to 5-29.

5-22 What about slowly lifting the mass from a table?
(from 5-11)

Go to 5-12.

5-23 The total energy of a system will not be conserved if there are any external forces. It
(from 5-12) must be explicitly stated that conservation laws refer only to closed systems.

Go to 5-24.

5-24 For a closed system the sum of the kinetic plus potential energy will always remain con-
(from 5-12 or 5-23) stant [*true* ; *false*] . 44

If incorrect go to 5-17 ; if correct go to 5-13.

5-25
(from 5-14)

The kinetic energy of block A just after the collision will be $\frac{1}{2}MV_A^2$ where

$V_A = $ _____ V_o .   **77**

Since $\left(\frac{1}{2}V_o\right)^2 = \frac{1}{4}V_o^2$, we have that the KE of block $A = \frac{1}{4}\left(\frac{1}{2}MV_o^2\right)$.

Then the kinetic energy of the two blocks together will be $\frac{1}{2}\left(\frac{1}{2}MV_o^2\right)$ [*true* ; *false*] .   **19**

Go to 5-15.

5-26
(from 5-34)

The potential energy of an object is equal to the work done on the object [by ; in opposing] a [conservative ; non-conservative] force. The potential energy [increases ; decreases] as the object is moved in the direction of this force.

83
21
10

Go to 5-35.

5-27
(from 5-15)

An amusement park ride consists of a car of mass M mass held by an arm of length L. The car is released from the top position while at rest. When the car passes through the horizontal position the [increase ; decrease] in its potential energy will be _____ .

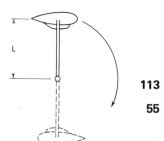

113
55

At the horizontal position its kinetic energy will have the value $\left[MgL \; ; \; \frac{1}{2}MgL \; ; \; \frac{1}{\sqrt{2}} \right]$ **35**

When the car passes through the bottom position its kinetic energy will have the value $\left[\frac{1}{2}MgL \; ; \; MgL \; ; \; 2MgL \; ; \; \text{none of these} \right]$. At this position its velocity will be [$\sqrt{gL}$; $\sqrt{2gL}$; $\sqrt{4gL}$; none of these] and its centripetal acceleration will be [g ; $2g$; $3g$; $4g$; none of these] .

4
97
134

At the bottom position its net acceleration will be _____ .   **59**

When passing through the bottom position the net force on a man of mass m will be _____ mg.   **28**

Go to 5-28.

5-28 At the bottom position the man's apparent weight will be _____ times his normal   66
(from weight.
5-27)

   If incorrect go to 5-16 ; if correct go to 5-30.

5-29 The correct statement should be: "Mass energy can be converted into other forms of en-
(from ergy." [*true* ; *false*]   73
5-21)

   Go to 5-20.

5-30 A mass M is moving in SHM. Its potential energy U
(from vs. position x is plotted at the right. If its total en-
5-28 or
5-16) ergy is W_o, at position x_1 its kinetic energy is
   [a ; b ; $(a+b)$; $(a-b)$] and its potential   31
   energy is [a ; b ; $(a+b)$; $(a-b)$] . The   106
   total mechanical energy is [a ; b ; $(a+b)$;   25
   $(a-b)$] . In this discussion assume the
   quantities a, b, and c are all energy magnitudes (always positive numbers).
   At position x_o the KE is _____ and the potential energy is _____ . At position x_o 70
                                                                                                               89
   the total mechanical energy is _____ .   37

   Go to 5-31.

5-31
(from 5-30)

The position from which zero energy is measured is arbitrary. Suppose it is convenient to have $U = 0$ when $x = x_2$. Then the previous potential diagram would look as follows (a, b, and c are still to be positive).

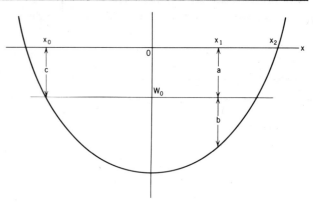

Now the total energy W_0 is [positive ; negative] . At position x_1 the KE is [a ; b ; $(a+b)$; $(a-b)$] , the potential energy is [a ; $-a$; b ; $-b$; $(a+b)$; $-(a+b)$] , and the total energy is [a ; $-a$; $-b$; $(a+b)$; $-(a+b)$] .

At position x_0 the KE is [0 ; c ; $-c$] , the potential energy is [0 ; c ; $-c$] , and the total energy is [0 ; c ; $-c$] .

139
8
46
30
16
86
112

Go to 5-18.

5-32
(from 5-19)

The condition for circular orbit is $\dfrac{V_c^2}{R} =$ [g ; $2g$; neither]

The KE while in circular orbit is $\dfrac{1}{2}mV_c^2$ or $\left[\dfrac{1}{2}mgR \; ; \; mgR \; ; \; 2mgR \; ; \; \sqrt{2}mgR \right]$.

The KE necessary for escape is $\dfrac{1}{2}mV_0^2$ which is [mgR^2 ; mgR ; $2mgR$; none of these] .

This last answer is _____ times the preceding answer.

50
7
67
34

Go to 5-21.

5-33
(from 5-20)

An antiproton is an anti-heavy particle. In other words a proton plus an antiproton is a total of zero heavy particles. Hence in counting heavy particles we count each proton or neutron as $+1$ and each antiproton or antineutron as -1.

An antiproton $\bar{P}$ can be produced by bombarding a proton with a high energy proton:

$$P + P \longrightarrow P + P + \bar{P} + ?$$

In the above reaction the question mark represents [no particle ; a proton ; an antiproton] . Ignoring the question mark, the total number of heavy particles on the right hand side is _____ . According to the law of conservation of heavy particles, the number of heavy particles on the right hand side must be _____ .

88

105

141

Go to 5-34.

5-34
(from 5-33)

The total number of protons and neutrons minus the total number of antiprotons and antineutrons in the universe must remain constant [*true* ; *false*] .

96

Go to 5-26.

5-35
(from 5-26)

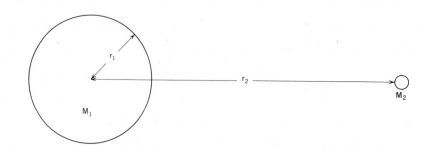

Consider a large spherical mass M_1 of radius r_1 and a small mass M_2 at a distance r_2 from the center of the sphere. The potential energy difference is $U_2 - U_1 =$

$\left[GM_1M_2 \dfrac{1}{r_1} - \dfrac{1}{r_2} \; ; \; GM_1M_2 \dfrac{1}{r_1} - \dfrac{1}{r_2} \; ; \; M_1G(r_2 - r_1) \; ; \; \text{neither} \right]$.   127

This is the work required to move M_2 from [r_1 to r_2 ; r_2 to r_1 ; ∞ to r_2 ; r_2 to ∞ ;   65
∞ to r_1] .

$(U_2 - U_1)$ is a [positive ; negative] quantity.   3

Go to 5-36.

5-36
(from 5-35)

When a charged particle q is placed between two charged parallel plates as shown, there will be an electric force F on it pointing to the right. If we are to do positive work on q, we must move q from plate [1 to 2 ; 2 to 1] .   39

The charge q will have a [higher ; lower] potential energy when at plate 1 than when at plate 2. The potential energy difference between the two plates is $\left[Fx \; ; \; \dfrac{Fx}{q} \; ; \; qFx \; ; \; \dfrac{q}{x} \; ; \; \text{none of these} \right]$.   118, 101

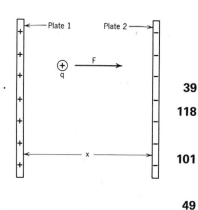

If q is free to move it will move toward plate [1 ; 2] .   49

Then its potential energy will [increase ; decrease] and its kinetic energy will   81
[increase ; decrease] .   18

Go to 5-52.

5-37
(from 5-4 or 5-8)

This frame begins a discussion of center of mass and statics. If you have been instructed to skip these subjects, go on to frame 5-10.

Suppose we have 3 different masses M_1, M_2, and M_3 located at $x_1, x_2,$ and x_3. The center of mass will then be located at $x_c = $

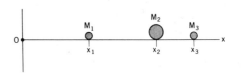

_____ .

85

Consider another observer whose origin is at O'. He will measure $x'_1 = x_1 + X$, $x'_2 = x_2 + X$ and $x'_3 = x_3 + X$.

He would say that the center of mass is at $x'_c = $ _____ where $M = M_1 + M_2 + M_3$.

123

Go to 5-38.

5-38
(from 5-37)

The observer measuring from O' would say that the center of mass is at

$$x'_c = \frac{(x_1 + X)M_1 + (x_2 + X)M_2 + (x_3 + X)M_3}{M}$$ [*true* ; *false*] .

107

This equation can be simplified by collecting the X terms:

$$x'_c = \frac{X(M_1 + M_2 + M_3)}{M} + \frac{x_1 M_1 + x_2 M_2 + x_3 M_3}{M}$$

The second term on the right hand side is the center of mass as measured from the origin O [*true* ; *false*] . The first term is X [*true* ; *false*] .

15
72

Go to 5-39

5-39
(from 5-38)

The final result of the preceding problem is that $x'_c = X + x_c$. Hence both observers predict the same position in space for the center of mass. Therefore the location of the center of mass is independent of the point from which it is measured.

Question: If an observer is sitting on the center of mass of a closed system, the total momentum of the system must always appear to be zero [*true* ; *false*] .

138

Go to 5-40.

5-40
(from 5-39)
If the sum of the forces on a rigid body is zero, the body cannot move [*true* ; *false*] .   111

If incorrect go to 5-44 ; if correct go to 5-49.

5-41
(from 5-49 or 5-50)
A condition for static equilibrium of a rigid body is that the sum of all torques on the body be zero [*true* ; *false*] .   63

Some physics textbooks imply that this is an independent fundamental law of nature; i.e., it does not follow directly from other more basic laws of nature.

All the conditions for static equilibrium can be deduced directly from Newton's laws of motion [*true* ; *false*] .   93

Go to 5-42.

5-42
(from 5-41)
A force F is applied to point B on a rigid body. The torque with respect to point A is _____ .   133

The torque with respect to point B is [O ; $R_1 F$; $R_2 F$; F ; none of these] .   109

Go to 5-45.

5-43
(from 5-45)
The sum of the torques on the rigid rod with respect to point A is _____ .   76

The sum of the torques with respect to point C is _____ .   5

If incorrect go to 5-46 ; if correct go to 5-47.

5-44
(from 5-40)
You have seen the previous frame several times before in Chapter 3. *Remember*, according to Newton's First Law a body can move with constant velocity even though $F_{net} = 0$.

Go to 5-49.

5-45
(from 5-42)

The torque about point A is defined as $T = R_1 \times F_\perp$
[true ; false] .
Remember, by using similar triangles it can be shown that $R_1 F_\perp = R_2 F$ (see frame 5-7).

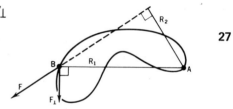

27

Go to 5-43.

5-46
(from 5-43)

The torque with respect to point C is _____ .

In the second figure the torque with respect to point C is _____ .

Now add the above two results:

Sum of torques = _____ .

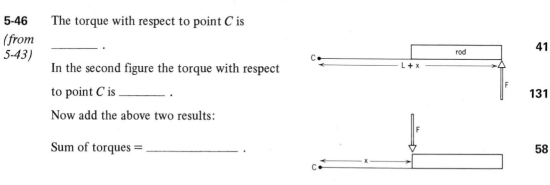

41

131

58

Go to 5-47.

5-47
(from 5-43 or 5-46)

A ladder of length L and mass M leans at an angle θ against a slippery wall. We wish to find the strength of the frictional force of the ground that pushes back on the ladder and prevents it from slipping. This frictional force is _____ .

If we count F_v and F_h as separate forces, there are _____ torques tending to rotate the ladder at about point B.

There are _____ torques tending to rotate the ladder about point A.

There are _____ torques tending to rotate the ladder about the center of mass.

Which point or points can be used as reference points for the torques? _____ .

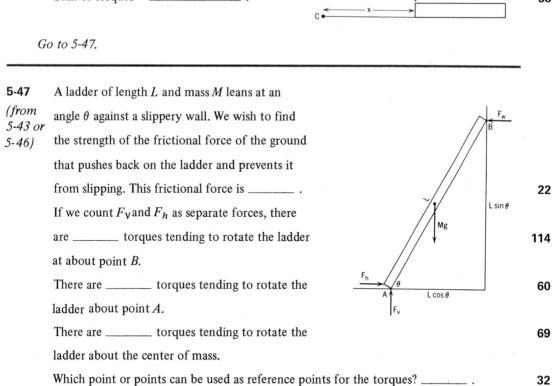

22

114

60

69

32

If incorrect go to 5-51 ; if correct go to 5-48.

5-48
(from 5-47 or 5-51)

Whichever point we select for the reference point, we will be able to solve the problem and get the correct answer. Let us arbitrarily choose B for the reference point. The torque due to F_h is then _____ .   98

The torque due to Mg is then _____ .   135

We can write down 3 simultaneous equations corresponding to the 3 conditions:

   a. Sum of vertical forces must be zero.

   b. Sum of horizontal forces must be zero.

   c. Sum of torques must be zero.

Condition _____ gives $F_h = F_w$ 29

Condition _____ gives $F_v = Mg$ 74

Condition _____ gives $F_h L \sin\theta + \frac{1}{2}MgL \cos\theta - F_v L \cos\theta = 0$.   36

Now solve this third equation for F_h. The result is

$$F_h = \frac{F_v L \cos\theta - \frac{1}{2}MgL \cos\theta}{L \sin\theta} .$$

Now eliminate F_v by substitution and obtain the final solution:

$F_h = $ _____ .   120

The answer is independent of L [*true* ; *false*] .   54

When $\theta = 45°$, $F_h = $ _____ .   24

Go to 5-10.

5-49
(from 5-40 or 5-44)

If a rigid body is initially at rest, it will stay completely motionless as long as the sum of the forces on it are zero; i.e., as long as $F_{net} = 0$ [*true* ; *false*] .   45

If incorrect go to 5-50 ; if correct go to 5-41.

5-50 If $F_{net} = 0$, then the sum of the torques must also always be zero [*true* ; *false*] .   129
(from
5-49) If the sum of the torques on a rigid body is not zero, the body must start rotating if initially at rest [*true* ; *false*] .   125
(Hint: consider spinning a penny.)

Go to 5-41.

5-51 For a rigid body in static equilibrium, if the sum of the torques about one axis is zero,
(from
5-47) then the sum of the torques about any other axis will also be zero [*true* ; *false*].   56
(Hint: see frame 5-43.)

Go to 5-48.

5-52 Review Questions:
(from
5-36)
1. List 4 conservation laws: 1. _____

2. _____

3. _____

4. _____

2. An airplane is doing loop-the-loop at a velocity $v = \sqrt{gR}$ where R is the radius of the loop.
   a. What is the acceleration of the plane? a = _____ g.
   b. What is the apparent weight of the pilot when at the bottom of the loop?
      F_w = _____ .
   c. What is the apparent weight of the pilot when at the top of the loop?
      F_w = _____ .
3. If a low-flying earth satellite has a kinetic energy of 10^{11} joules, how much more energy would it need in order to escape from the earth? _____ joules.

Chapter 5 Answer Sheet

Item Number	Answer	Item Number	Answer	Item Number	Answer
1	false	48	false	100	true
2	$\frac{1}{4}MV_0^2$	49	2	101	Fx
		50	g	102	a
3	positive	51	4	103	$\frac{1}{2}V_0$
4	$2\,MgL$	52	clockwise		
5	LF	53	the same	104	$\frac{1}{4}MV_0^2$
6	counterclockwise	54	true		
7	$\frac{1}{2}mgR$	55	MgL	105	1
		56	true	106	b
8	b	57	true	107	true
9	1	58	LF	108	the same as
10	decreases	59	$4g$	109	zero
11	true	60	2		
12	true	61	false	110	$\frac{1}{2}MV_0^2 - 2\,Mgh'$
13	$\frac{1}{2}$	62	mgR	111	false
		63	true	112	$-c$
14	true	64	lost	113	decrease
15	true	65	r_1 to r_2	114	3
16	zero	66	5	115	true
17	$\dfrac{R_1}{R_2}$	67	mgR	116	5
		68	remain the same	117	true
18	increase	69	3	118	higher
19	true	70	zero	119	$\dfrac{V_0^2}{2g}$
20	false	71	$\dfrac{V_0^2}{8g}$		
21	conservative	72	true	120	$Mg\,\dfrac{\cos\theta}{2\sin\theta}$
22	F_h	73	false	121	MV_0
23	true	74	a	122	heat energy
24	$\frac{1}{2}Mg$	75	true	123	$\dfrac{M_1 x_1' + M_2 x_2' + M_3 x_3'}{M}$
25	$(a+b)$	76	LF		
26	$\frac{1}{4}MV_0^2$	77	$\frac{1}{2}$	124	true
		78	neither	125	true
27	true	79	gm cm/sec^2	126	$\frac{1}{4}$
28	4	80	clockwise		
29	b	81	decrease	127	$GM_1 M_2\left(\dfrac{1}{r_1} - \dfrac{1}{r_2}\right)$
30	$-a$	82	true		
31	a	83	in opposing	128	$\frac{1}{2}MV_0^2$
32	any point	84	false	129	false
33	neither	85	$\dfrac{M_1 x_1 + M_2 x_2 + M_3 x_3}{M_1 + M_2 + M_3}$	130	false
34	2			131	$-xF$
35	MgL	86	$-c$	132	greater
36	c	87	2	133	$R_2 F$
37	$(a+b)$	88	a proton	134	$4g$
38	$(2M)gh'$	89	c		
39	2 to 1	90	similar	135	$\frac{1}{2}MgL\cos\theta$
40	$\frac{1}{2}mV_0^2$	91	same	136	true
		92	Mgh	137	true
41	$(L+x)F$	93	true	138	true
42	$MR_2 V\perp$	94	true	139	negative
43	less than	95	clockwise		
44	false	96	true	140	$\frac{1}{2}MV_0^2$
45	false	97	$\sqrt{4gL}$		
46	$-(a+b)$	98	$F_h L\sin\theta$	141	2
47	true	99	same as	142	external

Chapter 6
THE KINETIC THEORY

6-1 Intergalactic space contains a low density of hydrogen atoms. The density is $D_0 \approx 10^{-30}$ gm/cm³ (the symbol $\approx$ means approximately equal). In this problem we wish to estimate the average distance between the hydrogen atoms in outer space. If M is the mass of a single hydrogen atom, the average volume occupied by each hydrogen atom is

$V_0 = \left[D_0 M \; ; \; \dfrac{M}{D_0} \; ; \; \dfrac{D_0}{M} \; ; \; \text{none of these} \right]$. The side of a cube having volume V_0 is   **76**

$L = \underline{\qquad}$.   **45**

The approximate value for M is $M \approx 10^{-24}$ gm per hydrogen atom.

Then $V_0 \approx \underline{\qquad}$ cm³ and $L_0 \approx \underline{\qquad}$ cm.   **55**
   16

Go to 6-2.

6-2 The density of gas in a box is D. The gas consists of N molecules of mass m each.
(from 6-1) The number of molecules in the box is $N = \underline{\qquad}$.   **23**

If incorrect go to 6-11 ; if correct go to 6-3.

6-3
(from 6-2 or 6-11)

A gas is confined in a cylinder. It takes a force F to push the piston a distance ΔL. If the area of the piston is A cm², the work done on the gas in compressing it a distance L is $W = \left[\dfrac{F \Delta L}{A} \ ; \ FA \ \Delta L \ ; \ F \Delta L \ ; \ \text{none of these} \right]$.

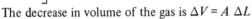

The decrease in volume of the gas is $\Delta V = A \ \Delta L$
[*true* ; *false*] .   **36**

Hence the work done on the gas is $W = \left[P \Delta V \ ; \ \dfrac{P \Delta V}{A} \ ; \ P \Delta L \ ; \ PA \ \Delta V \right]$.   **67**

Go to 6-4.

6-4
(from 6-3)

If a volume of gas in a cylinder is allowed to push the piston and expand its volume by ΔV, the work done by the gas is   [$P \Delta V$; $-F \Delta V$; neither] .   **42**

Pressure has the same units as [energy ; angular momentum ; energy per unit volume ; none of these] .   **10**

Go to 6-5.

6-5
(from 6-4)

The pressure at the bottom of a beaker of mercury is measured to be $1\tfrac{1}{2}$ atmospheres. If the beaker of mercury is placed in an evacuated vessel, the pressure at the bottom of the beaker will be [greater ; less ; the same] .   **21**

Go to 6-6.

6-6
(from 6-5)

If a beaker of mercury is placed in an evacuated vessel the pressure in the mercury will be reduced by $\left[0 \ ; \ \dfrac{1}{2} \ ; \ 1 \ ; \ 1\dfrac{1}{2} \right]$ atmospheres.   **75**

Go to 6-7.

6-7
(from 6-6)
If the pressure at the bottom of a beaker of mercury is $1\frac{1}{2}$ atmospheres when out in the open, the height of the mercury in the beaker is _____ cm. (A pressure of 1 atm. reads as a height of 76 cm on a mercury barometer.)

69

If incorrect go to 6-17 ; if correct go to 6-8.

6-8
(from 6-7)
An automobile tire has a gauge pressure of 2 atmospheres. (When deflated the gauge pressure is zero.) The absolute pressure inside the tire is _____ atmospheres. The pressure just outside the tire is _____ atmospheres.

4

33

Go to 6-9.

6-9
(from 6-8)
A crate of mass M_1 and volume V_1 is placed on board a ship. The mean density of the crate is then $D_1 = \dfrac{M_1}{V_1}$. This will cause the waterline on the hull of the ship to

[rise; lower ; remain the same] . If the waterline rises, the ship displaces more water [*true ; false*] .

71

46

Go to 6-10.

6-10
(from 6-9)
The additional water displaced by the ship will have a volume [0 ; V_1; neither] . The additional water displaced by the ship will have a mass

$\left[M_1 \ ; \ D_1 M_1 \ ; \ \dfrac{M_1}{D_1} \ ; \text{ none of these} \right]$. The answer depends on whether D_1 is greater than D_w the density of the water [*true ; false*] .

12

19

59

Go to 6-12.

6-11
(from 6-2)
The total mass in the box is Nm [*true ; false*].

The density of the gas is $D = \dfrac{M}{V}$ where $M = \left[m \ ; \ \dfrac{m}{N} \ ; \ Nm \ ; \text{ none of these} \right]$.

20

40

Go to 6-3.

6-12
(from 6-10)

The volume of the additional water displaced is $\left[\dfrac{M_1}{D_1} \; ; \; \dfrac{M_1}{D_w} \; ; \; M_1 \dfrac{D_w}{D_1} \; ; \; \dfrac{M_1}{D_1 D_w} \; ; \right.$ none of these $\left.\right]$.

61

If incorrect go to 6-21 ; if correct go to 6-13.

6-13
(from 6-12 or 6-21)

Archimedes' principle only applies to bodies floating in fluids [*true* ; *false*].

26

Boyles' law states that the product $P_1 V_1$ for any one gas equals the product $P_2 V_2$ for any other gas [*true* ; *false*].

13

Go to 6-14.

6-14
(from 6-13)

Boyle's law states that if both gases are at the same temperature then $P_1 V_1$ of one gas equals $P_2 V_2$ of the second gas [*true* ; *false*].

7

Go to 6-18.

6-15
(from 6-22)

The quantity $P_1 V_1 = \left[NkT \; ; \; \dfrac{3}{2} NkT \; ; \; NmkT \right]$ for N molecules of a perfect gas at temperature T.

73

A second perfect gas containing N molecules and at the same temperature would have $P_2 V_2$ equal to the same quantity [*true* ; *false*].

48

Go to 6-16.

6-16
(from 6-22 or 6-15)

If gas 1 and gas 2 are at the same temperature, $\overline{v_1^2}$ will equal $\overline{v_2^2}$ [*true* ; *false*].

52

Go to 6-19.

6-17
(from 6-7)
With only air in the beaker, the pressure reading would correspond to _____ cm of mercury on a barometer.

With 76 cm of mercury in the beaker, the pressure reading at the bottom would be _____ atmospheres when out in the open. With 38 cm of mercury in the beaker, the pressure reading at the bottom of the mercury would be _____ atmospheres when out in the open.

30

57

2

Go to 6-8.

6-18
(from 6-14)
If M grams of one perfect gas is at the same temperature as M grams of another perfect gas, then P_1V_1 will always equal P_2V_2 [*true* ; *false*].

63

If incorrect go to 6-36 ; if correct go to 6-22.

6-19
(from 6-16)
When two different gases are in thermal equilibrium the average momentum per molecule of gas 1 will equal the average momentum per molecule of gas 2 [*true* ; *false*]. The previous statement is correct if the word momentum is replaced by translational kinetic energy [*true* ; *false*].

43

72

Go to 6-20.

6-20
(from 6-19)
Consider N molecules of a perfect gas at a temperature T in a volume V. The total translational kinetic energy of the molecules is $\left[PV \ ; \ \dfrac{PV}{T} \ ; \ kT \ ; \ NkT \ ; \ \dfrac{3}{2}NkT \ ; \ \text{none of these} \right]$.

If the temperature is increased by 1°C the increase in the translational kinetic energy of this sample of gas will be $\left[0 \ ; \ NkT \ ; \ Nk \ ; \ \dfrac{3}{2}Nk \ ; \ PV \ ; \ \text{none of these} \right]$.

14

38

Go to 6-23.

6-21
(from 6-12)

Let V_w be the volume of the water displaced (it has mass M_1). Then
$$D_w = \left[\frac{V_w}{M_1} \ ; \ M_1 V_w \ ; \ \frac{M_1}{V_w} \right].$$

28

Go to 6-13.

6-22
(from 6-18 or 6-36)

If N molecules of one perfect gas are at the same temperature as N molecules of another perfect gas, then $P_1 V_1 = P_2 V_2$ [*true* ; *false*].

50

If incorrect go to 6-15 ; if correct go to 6-16.

6-23
(from 6-20)

The specific heat C_v of a gas is defined as the amount of heat energy that must be given to a mole of gas at constant volume to raise its temperature by 1°C. For a perfect monatomic gas $C_v = \left[\frac{3}{2} N_o k \ ; \ N_o k \ ; \ \frac{5}{2} N_o k \ ; \ \text{none of these} \right]$ where $N_o = 6.02 \times 10^{23}$ and $k = 1.38 \times 10^{-16}$ erg/degree.

6

If 1 calorie of heat is 4.18×10^7 ergs of energy, what is C_v in cal/mole degree?

$C_v =$ _____ cal/mole degree.

66

Go to 6-24.

6-24
(from 6-23)

In addition to the translational kinetic energy, diatomic molecules have 2 additional rotational degrees of freedom at room temperature. The rotational kinetic energy of a diatomic gas is _____ times the translational kinetic energy.

11

Go to 6-26.

6-25
(from 6-39)

The absolute temperature of a perfect gas is given by $T = \frac{2}{3}\frac{KE}{k}$ where KE is the average translational KE per molecule [*true* ; *false*]. This formula is independent of the density of the particles in the gas [*true* ; *false*].

18

39

Go to 6-40.

6-26
(from 6-24)

The total kinetic energy content of a diatomic gas sample containing N molecules is _____ times NkT.

The specific heat C_v for a diatomic gas would be $C_v = $ _____ .

The ratio of the specific heats for a diatomic to a monatomic gas would be _____ .

65

32

9

Go to 6-27.

6-27
(from 6-26)

The temperature of a fixed volume of a perfect gas is increased by 1°C. The fractional increase in pressure would be $\left[\frac{1}{T} ; \frac{5}{3T} ; \frac{3}{2T} ;$ none of these $\right]$.

15

If incorrect go to 6-33 ; if correct go to 6-28.

6-28
(from 6-27 or 6-33)

There is a second kind of specific heat C_p defined as the amount of heat that must be put into 1 mole of gas kept at constant pressure in order to raise the temperature by 1°C. Now if P is kept fixed and T is increased, the volume V will [increase ; decrease ; remain the same] .

We would have the relation $P \Delta V = N_0 k \Delta T$ in pushing a piston [*true* ; *false*] .

The gas would do an amount of mechanical work $P\Delta V$ in pushing a piston [*true* ; *false*] .

The gas would do an amount of work $N_0 k \Delta T$ in pushing a piston [*true* ; *false*] .

This work done by the gas has the value $\left[N_0 k ; \frac{3}{2} N_0 k ; N_0 kT ;$ none of these $\right]$ each time the temperature is increased by one degree.

37

25

49

70

54

Go to 6-29.

79

6-29
(from 6-28)

The total energy that must be supplied to a mole of gas kept at constant pressure in order to raise its temperature by 1°C is [C_v ; C_vT ; $C_v + N_0k$; $C_v + N_0kT$; none of these] . 74

The difference between two kinds of specific heats is $C_p - C_v =$ [0 ; N_0kT ; N_0k ; $\frac{3}{2}N_0k$; none of these] . 77

The quantity $C_p - C_v$ has a value of about _____ calories/mole degree. 35

Go to 6-30.

6-30
(from 6-29)

Classical physics using Newton's laws makes definite predictions for the specific heats of gases [*true* ; *false*]. 17

Experimentally the specific heats of gases are usually [lower than ; higher than ; the same as] that predicted by classical physics. 41

Go to 6-31.

6-31
(from 6-30)

Hydrogen gas H_2 has a molecular weight of _____ . 3

Oxygen gas O_2 has a molecular weight of _____ . 68

Water H_2O has a molecular weight of _____ . 29

One mole of water contains [N_0 ; $2N_0$; $3N_0$; $4N_0$; none of these] molecules. 47

One mole of water has a mass of _____ gms. 60

Go to 6-32.

6-32
(from 6-31)

The mass of a water molecule is $\frac{18 \text{ gm}}{N_0}$ [*true* ; *false*]. 8

Go to 6-34.

6-33
(from 6-27)

We can obtain the fractional increase in pressure ΔP from the equation $P = \dfrac{NkT}{V}$.

Then ΔP would equal $\dfrac{Nk\Delta T}{V}$ [*true* ; *false*].   58

Now divide this expression for ΔP by the expression for P. The result is $\dfrac{\Delta P}{P} = \dfrac{\Delta T}{T}$

[*true* ; *false*].   31

Go to 6-28.

6-34
(from 6-32)

One cubic centimeter of a liquid is boiling at a pressure P_1 and temperature T_1. The vapor pressure curve of this liquid is shown in the figure. If the pressure is held at $P = P_1$ and the temperature is increased, [*all* ; *some* ; *none*] of the liquid will become gas or vapor.

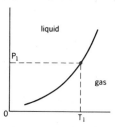

    5

If the temperature is kept at $T = T_1$ and the pressure is reduced, [*all* ; *some* ; *none*] of the liquid will become gas or vapor.   24

In order to maintain the temperature T_1 while P is reduced from P_1, heat must flow [*into* ; *out of* ; *neither*] the liquid.   44

Go to 6-35.

6-35
(from 6-34)

The latent heat of vaporization of H_2O at $T = 100°C$ is 540 calories per gram. Suppose we have 1 gm of liquid H_2O at $T = 100°C$ and the pressure is maintained at one atmosphere. Then 1 calorie of heat is added to this quantity of H_2O. The temperature of the H_2O will then [*increase* ; *decrease* ; *remain the same*].   62

If incorrect go to 6-37 ; if correct go to 6-38.

6-36
(from 6-18)

Boyles' law as stated applies to a fixed quantity of just one gas sample under different pressure and volume conditions, but the same temperature [*true* ; *false*].   51

Go to 6-22.

6-37 The temperature of a kettle of boiling water will not increase beyond 100°C as heat is
(from added [*true* ; *false*].   34
6-35)
Steam can be heated to temperatures greater than 100°C [*true* ; *false*]. Water can   64
be heated to temperatures greater than 100°C and still remain in the liquid state if the
pressure is [greater than ; less than ; neither] one atmosphere.   22

Go to 6-38.

6-38 A machine designed to convert the heat energy of sea water into useful mechanical
(from work must violate the law of conservation of energy [*true* ; *false*]. Such a machine   53
6-35 or
6-37) must violate the second law of thermodynamics [*true* ; *false*].   27

Go to 6-39.

6-39 Consider a container of gas where the average kinetic energy per molecule is kept
(from constant. If the number of molecules in the container is increased the gas temperature
6-38)
will [increase ; decrease ; remain the same] .   56

If incorrect go to 6-25 ; if correct go to 6-40.

6-40 Review Questions:
(from 1. $C_p - C_v = N_0 k$ times _____ .
6-39 or
6-25) 2. For a monatomic gas $C_v = N_0 k$ times _____ .
3. For a diatomic gas $C_v = N_0 k$ times _____ .
4. One mole of CH_4 contains _____ gms of methane.
5. One mole of CH_4 contains _____ molecules.
6. What is the density of a perfect gas in terms of P, k, T, and m (the mass of each molecule)? $D = $ _____
7. Fill in the accompanying table of specific heats per mole of monatomic and diatomic perfect gases. Express the specific heats in terms of N_0 and k.

	Monatomic	Diatomic
C_v		
C_p		

Chapter 6 Answer Sheet

Item Number	Answer
1	$F\Delta L$
2	$1\frac{1}{2}$
3	2
4	3
5	all
6	$\frac{3}{2}N_0 k$
7	false
8	true
9	$\frac{5}{3}$
10	energy per unit volume
11	$\frac{2}{3}$
12	neither
13	false
14	$\frac{3}{2}NkT$
15	$\frac{1}{T}$
16	100
17	true
18	true
19	M_1
20	true
21	less
22	greater than
23	$\frac{DV}{m}$
24	all
25	true
26	false
27	true
28	$\frac{M_1}{V_w}$
29	18
30	76
31	true
32	$\frac{5}{2}N_0 k$
33	1
34	true
35	2
36	true

Item Number	Answer
37	increase
38	$\frac{3}{2}Nk$
39	true
40	Nm
41	lower than
42	$P\Delta V$
43	false
44	into
45	$V_0^{\frac{1}{3}}$
46	true
47	N_0
48	true
49	true
50	true
51	true
52	false
53	false
54	$N_0 k$
55	10^6
56	remain the same
57	2
58	true
59	false
60	18
61	$\frac{M_1}{D_w}$
62	remain the same
63	false
64	true
65	$\frac{5}{2}$
66	3
67	$P\Delta V$
68	32
69	38
70	true
71	rise
72	true
73	NkT
74	$C_V + N_0 k$
75	1
76	$\frac{M}{D_0}$
77	$N_0 k$

Chapter 7
ELECTROSTATICS

7-1 The element beryllium has an atomic "weight" of 9 and atomic number Z = 4. Hence the Be nucleus contains _____ protons and _____ neutrons. The neutral Be atom has _____ orbital electrons of [positive ; negative] charge.

38
75
4
46

Go to 7-2.

7-2
(from 7-1)
The net electric charge of an object is proportional to its mass. [*true* ; *false*].

A metal ball will be slightly [lighter ; heavier] when it is negatively charged compared to when it is positively charged.

108
59

Go to 7-4.

7-3
(from 7-4)
In the CGS system Coulomb's law is $F = Q_1 Q_2 / r^2$. Hence the force between two charges of 3×10^9 esu separated by 1 meter is $F = \left[(3 \times 10^9)^2 \; ; \; \dfrac{(3 \times 10^9)^2}{100^2} \; ; \; 3 \times 10^9 \; ; \; \text{neither} \right]$ dynes.

95

Go to 7-5.

7-4
(from 7-2)
In the MKS system F is measured in _____ , r in meters, and Q in _____ . We wish to find K in the MKS version of Coulomb's law where $F = K Q_1 Q_2 / r^2$. First, what is the force in dynes between two charges of 1 coulomb each separated by 1 meter? If we remember that 1 coulomb equals _____ statcoulombs we can use Coulomb's law in CGS units (also called esu units) to obtain

$F =$ _____ dynes.

30
58
139
87

If incorrect go to 7-3 ; if correct go to 7-5.

7-5   The force between two charges of 1 coulomb each separated by 1 meter is _____   60
(from   newtons. In the MKS equation $F = KQ^2/r^2$ $F =$ _____ MKS units when Q and r are   15
7-4 or
7-3)   both 1 MKS unit each.

Numerically we have $(9 \times 10^9) = K(1)^2/(1)^2$. Hence K must be _____ . The units   52
of K are _____ .   22

Go to 7-6.

7-6   In the CGS version of Coulomb's law between identical charges we can solve for Q and
(from   obtain equivalent mechanical units. Solving $F = Q^2/r^2$ for Q gives $Q =$ _____ .   49
7-5)
Hence the equivalent mechanical units for Q would be _____ .   96

Go to 7-7.

7-7   Most substances contain atoms whose nuclei are composed of equal numbers of protons
(from   and neutrons. If the mass of the proton and neutron is 1.66×10^{-24} gm each, one gram
7-6)   of such a substance will contain _____ electrons.   113

If incorrect go to 7-14 ; if correct go to 7-8.

7-8   Suppose one out of each billion electrons is removed from this gram. Then the number of
(from   missing electrons will be _____ .   10
7-7 or
7-14)   The net charge in statcoulombs on this gram of material will be _____ times the   78
             previous answer. The result is a [positive ; negative] charge of _____   155
             statcoul.   32

Go to 7-11.

7-9   Let M_p be the mass of the proton and m the mass of the electron. The ratio of the
(from   electrostatic force between a proton and electron to the gravitational force is
7-11)
             _____ .   125

If incorrect go to 7-15 ; if correct go to 7-10.

7-10
(from 7-9 or 7-15)

An electroscope has zero excess charge. The two leaves will be [deflected apart ; hanging together] . A positively charged rod is brought near the electroscope. The leaves now contain [negative ; positive ; zero] excess charge. The [electrons ; protons] in the [metal ball; leaves] are [repelled ; attracted] into the [ball ; leaves] .

29
50
83
100
23
1

Go to 7-16.

7-11
(from 7-8)

In the Bohr model of the hydrogen atom the electron is attracted to the proton by the electrostatic force. Suppose the only force between the electron and proton were the attractive gravitational force. Then what would be the velocity of the electron (in a circular orbit of radius 0.53×10^{-8} cm)?

The gravitational force on the orbital electron will be $F = $ [$M_p m/r^2$; GM_p/r^2 ; $GM_p m/r^2$] where M_p is the mass of the proton and m the mass of the electron. F will also have the value [$M_p v^2/r$; mv^2/r ; mv^2/r^2 ; neither] . Equating the two previous results gives $v^2/r = GM_p/r^2$. Now solve for v and we obtain v = _____.

If we insert the CGS values into the above result we obtain the value $v = 4.5 \times 10^{-12}$ cm/sec. To obtain the period of revolution we divide [v by r ; v by the circumference ; the circumference by v] . The result is _____ sec. This is indeed a very sluggish hydrogen atom. The true period is _____ sec.

63
102
40
111
73
16

Go to 7-9.

7-12
(from 7-25)

The electric field at point P is _____ .

28

Go to 7-18.

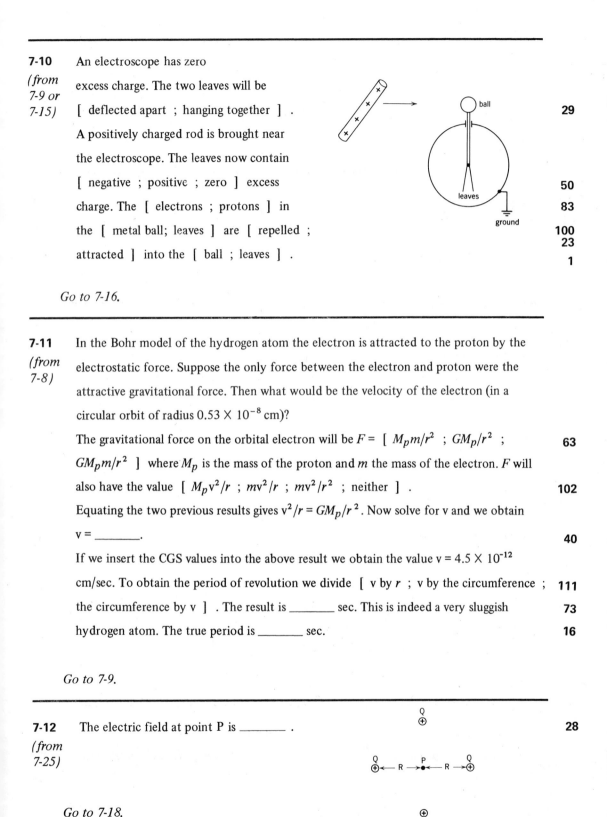

7-13
(from 7-20)

If the negatively charged leaves decrease their deflection additional electrons must [leave ; enter] the leaves. A [positive ; negative] rod would attract electrons from the leaves to the ball.

84
145

Go to 7-21

7-14
(from 7-7)

The number of protons per gram will be _____ times Avogadro's number. The number of electrons per gram will be [one half ; the same as ; twice] the number of protons per gram.

118
94

Go to 7-8.

7-15
(from 7-9)

The electrostatic force between proton and electron is _____ .

The gravitational force between proton and electron is _____ .

In taking the ratio of the two above results, the r^2 terms cancel out [*true* ; *false*].

27
115
48

Go to 7-10.

7-16
(from 7-10)

Consider an electroscope that initially has a positive excess charge. If a positively charged rod is brought near the electroscope, some electrons in the [leaves ; metal ball] will be [repelled ; attracted] into the [ball ; leaves] . The leaves will now contain [more ; less ; the same number of] electrons [than ; as] before and will consequently be deflected [more ; less ; the same] .

69
34
123
20
77
56

Go to 7-19.

7-17
(from 7-19 or 7-40)

A negatively charged rod is brought near an uncharged electroscope. The ball is touched by hand. Electrons will flow [into ; out of] the ball leaving the electroscope with a net [positive ; negative] charge.

104
105

Go to 7-20.

7-18
(from 7-12)
The electric field at point P is zero [*true* ; *false*].   133

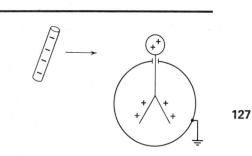

Go to 7-26.

7-19
(from 7-16)
A negatively charged rod is brought near a positively charged electroscope. The leaves will [come together ; separate more ; stay the same] .   127

If incorrect go to 7-40 ; if correct go to 7-17.

7-20
(from 7-17)
An electroscope has a net negative charge. A rod of unknown charge is brought nearby and the deflection of the leaves is observed to decrease. The charge on the rod must be [positive ; negative] .   158

If incorrect go to 7-13 ; if correct go to 7-21.

7-21
(from 7-20 or 7-13)
Any object containing a total charge Q will produce an electric field. $E =$ [Q/r ; Qq/r^2 ; Q/r^2 ; neither] .   154

If incorrect go to 7-45 ; if correct go to 7-22.

7-22
(from 7-21 or 7-45)
The electrostatic force between a point charge Q_1 and an extended body of total charge Q_2 will be $F =$ [$Q_1 Q_2 / r^2$; Q_1/r^2 ; neither] . The force will also have the value $F =$ [$Q_2 E$; $Q_1 E$; neither] where E is the electric field produced by [Q_1 ; Q_2 ; both Q_1 and Q_2] .   13
    42
    25

Go to 7-23.

7-23
(from 7-22)

The net field at point P will point [up ; down ; left ; right ; neither (it will be zero)] .

91

If incorrect go to 7-28 ; if correct go to 7-24.

7-24
(from 7-23 or 7-28)

The net electric field at point P will point [up ; down ; left ; right] . The strength of the electric field produced by $+Q$ alone will be $\dfrac{Q}{r'^2}$ where $r' = \left[\, r \;;\; r + \dfrac{L}{2} \;;\; r - \dfrac{L}{2} \;;\; \text{neither} \,\right]$.

Hence $E = \left[\, LQ/r^3 \;;\; Q/r^2 \;;\; \dfrac{Q}{r^2\left(+\dfrac{1}{2}L\right)^2} \;;\; \dfrac{Q}{\left(r+\dfrac{1}{2}L\right)^2} \;;\; \text{neither} \,\right]$ is the field produced by Q alone. The force on a unit positive charge at P due to $+Q$ alone will be [repulsive ; attractive] and pointing to the [left ; right] .

The field produced by the charge $-Q$ will be [larger ; smaller] in magnitude and will be pointing to the [left ; right] . The magnitude of the net field will be _____ .

Go to 7-25.

2
7
106
141
45
80
53
149

7-25
(from 7-24)

The net electric field at point P is
[Q/R^2 ; $2Q/R^2$; neither] .

117

Go to 7-12.

7-26
(from 7-18)

A cubical box whose sides are L cm long contains two point charges Q_1 and Q_2. The number of lines of force leaving the box depends on the positions of Q_1 and Q_2 [*true* ; *false*]. The number of lines of force leaving the box depends on the area of the box [*true* ; *false*]. The number of lines leaving the box will be _____ .

132
54
110

Go to 7-29.

7-27 Dielectrics have the property that when a slab is put in a
(from field E, an induced charge σ' appears on the surface which
7-44) is proportional to E. In this frame we wish to derive a
relation between σ' and the dielectric constant ϵ. The
definition of ϵ is that the reduced field inside the slab
is E/ϵ. The number of lines of E per cm^2
leaving the left hand plate is _____ . The field in   5
region I is the same as the above answer [*true* ; *false*
false] . The number of these lines passing through   37
the dielectric is _____ . The field inside   120
the dielectric would then be $E' = $ [$4\pi\sigma'$; $4\pi(\sigma - \sigma')$; neither] E' also have the value   157
[ϵE ; E/ϵ] which is $4\pi\sigma/\epsilon$.   130
Now equate the two above expressions for E' and solve for σ'. The result is
$\sigma' = $ _____ .   147

Go to 7-47.

7-28 Suppose both q and Q are positive
(from charges. Then the direction of force
7-23) on q will be [A ; B ; C ; D] .

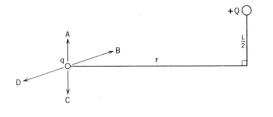

If q has the value [4.8×10^{-10} ; 1 ;   12
3×10^{10}] esu then this force on   36
q will be the value of the electric
field at q produced by [q ; Q] .   66

Go to 7-24.

7-29
(from 7-26)

A uniformly charged sphere of 10 cm radius has a total surface charge of 100 statcoulombs. The field at $r = 20$ cm is $E =$ _____ dynes per statcoul. The field at $r = 9$ cm is $E =$ _____ dyne/statcoul. The field at $r = 0$ is $E =$ _____ dyne/statcoul.

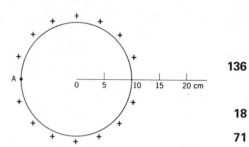

136

18

71

Now suppose an additional point charge of 10 statcoul. is attached to the surface at point A. Assume the initial uniform charge distribution of 100 statcoul. remains unchanged. Then the field at $r = 0$ would be $E =$ _____ dynes/statcoul.

89

Go to 7-30.

7-30
(from 7-29)

Consider three concentric charged spherical shells of radii R_1, R_2, R_3 and total charges Q_1, Q_2, Q_3. Point P is at a distance x from the surface of the outer sphere. The field at point P is _____ . The total number of lines traveling out to infinity would be [Q_3 ; $4\pi(Q_1 + Q_2 + Q_3)$; neither] .

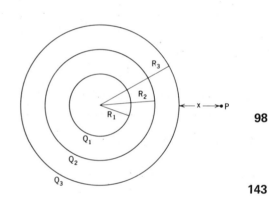

98

143

If incorrect go to 7-34 ; if correct go to 7-31.

7-31
(from 7-30 or 7-34)

An area of 3 cm² is at an angle of 30° to a uniform field of $E = 100$ dynes/statcoul. The number of lines of force cutting through the 3 cm² area will then be $N =$ _____ .

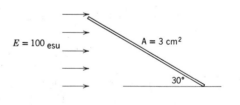

65

Go to 7-32.

7-32
(from 7-31)

Consider two parallel charged wires of charges ρ_1 and ρ_2 statcoul/cm respectively. We wish to calculate the force on a 1 cm length of Wire 2. This will be

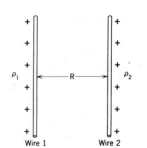

$F = \dfrac{\rho_1 \rho_2}{R^2}$ [*true* ; *false*].   11

The force on one cm of Wire 2 will be the same as the force on 1 cm of Wire 1 [*true* ; *false*] (except for sign).   103

Let E_1 be the field due to Wire 1 only. Then the force on 1 cm of Wire 2 will be

$F = \left[\rho_1 E_1 \; ; \; \rho_2 E_1 \; ; \; \dfrac{\rho_2 E_1}{R^2} \; ; \; \text{neither} \right]$.   51

The value of E_1 at the position of Wire 2 is $E_1 =$ _____ .   153

The force on a 1 cm length of either wire is _____ .   138

Go to 7-35.

7-33
(from 7-35)

Consider two point charges Q and q. The "absolute" potential energy is $U = qQ/r$. This is the amount of work required to move

[q to Q ; q from Q to its present position ;

q to ∞ ; q from ∞ to its present position] . U is [positive ; negative] if both   152
                                                                                                                                                44
charges are the same. The potential energy between a proton and electron is always

[positive ; negative] .   76

Go to 7-38.

7-34
(from 7-30)

Let us draw an imaginary sphere of radius $(R_3 + x)$ around the 3 inner spheres. The total charge contained in our imaginary sphere is _____ . The total number of lines of force cutting through our imaginary sphere is $N =$ _____ . The total area of our imaginary sphere is $A =$ _____ . The ratio N/A is

$$\frac{Q_1 + Q_2 + Q_3}{(x + R_3)^2} \quad [\ true\ ;\ false\ ].$$ This ratio is also the value of the electric field at point P [*true* ; *false*].

128

9

39

74

35

Go to 7-31.

7-35
(from 7-32)

Consider a sphere of radius r_a and charge Q and a point charge q at a distance r_b from the center of the sphere. Then the potential energy difference $U_b - U_a =$ $qQ(1/r_b - 1/r_a)$ [*true* ; *false*].

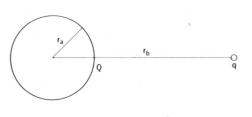

This is the work required to move q from [r_a to r_b ; r_b to r_a ; ∞ to r_b ; r_b to ∞] .
If Q and q are both positive $U_b - U_a$ is a [positive ; negative] quantity. If Q and q are of opposite charge the force will be attractive (as is the gravitational force) and $U_b - U_a$ will be [positive ; negative] .

101

134

79

17

Go to 7-33.

7-36
(from 7-42)

We saw in the previous frame that $\Delta U = -qE\Delta x$ [*true* ; *false*].
Hence $\Delta U/q = -E\Delta x$
    or $\Delta V = -E\Delta x$
  and $E = -\Delta V/\Delta x$.

151

Go to 7-37.

7-37
(from 7-42 or 7-36)

The potential at the surface of a charged spherical shell of total charge Q is

$V =$ _____ where R is the radius. 122

The potential at the center of the above shell is

$V =$ _____ . 93

If incorrect go to 7-43 ; if correct go to 7-44.

7-38
(from 7-33)

This is a plot of $U = -e^2/r$ which is the potential energy of an electron and proton separated by a distance r. In this diagram let r_0 represent the distance between electron and proton when their kinetic energy is zero. Then at any position r

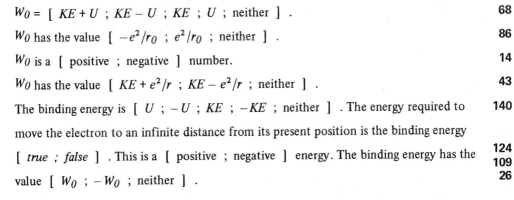

$W_0 = $ [$KE + U$; $KE - U$; KE ; U ; neither] . 68

W_0 has the value [$-e^2/r_0$; e^2/r_0 ; neither] . 86

W_0 is a [positive ; negative] number. 14

W_0 has the value [$KE + e^2/r$; $KE - e^2/r$; neither] . 43

The binding energy is [U ; $-U$; KE ; $-KE$; neither] . The energy required to 140
move the electron to an infinite distance from its present position is the binding energy
[*true* ; *false*] . This is a [positive ; negative] energy. The binding energy has the 124
value [W_0 ; $-W_0$; neither] . 109
26

Go to 7-39.

7-39
(from 7-38)

Suppose the electron has the total energy W_0 and is in a circular orbit. The radius of this must be [less than ; greater than ; the same as] r_0.

In this frame we will obtain a formula for this radius in terms of r_0. To do so we must first find the relation between W_0 and the radius of the circular orbit r_c.

In the circular orbit $W_0 = \frac{1}{2}mv^2 - e^2/r_c$ [true ; false]. Now we must express $\frac{1}{2}mv^2$ in terms of [r_0 ; r_c]. We can do this by equating the centripetal force to the [electric potential energy ; gravitational force ; electric force].

Hence $mv^2/r_c = $ [e^2/r_c ; e^2/r_c^2 ; neither]. Now multiply both sides by $\frac{1}{2}r_c$. The left hand side becomes the kinetic energy [true ; false]. The right hand side becomes _____ which is also $\left[\frac{1}{2}U \; ; -\frac{1}{2}U \; ; -U \; ; \text{neither} \right]$. Our equation $W_0 = \frac{1}{2}mv^2 - e^2/r_c$ becomes $W_0 = $ [$-e^2/2r_c$; 0 ; $-e^2/r_c$; $-2e^2/r_c$; neither].

From the previous frame we had a relation between W_0 and r_0:
$W_0 = $ [$-e^2/r_0$; $-e^2/2r_0$; neither]. Equating the last two results gives $r_c = $ _____ .

3

114

70

81

6

47

156
112

19
55
144

Go to 7-41.

7-40
(from 7-19)

When the negatively charged rod is brought near a positively charged electroscope, electrons in the [ball ; leaves] will be [attracted ; repelled] into the [ball ; leaves] thus making the leaves [more ; less] positive. The deflection of the positively charged leaves will then [increase ; decrease].

135
24
85
159
41

Go to 7-17.

7-41
(from 7-39)
There is a uniform electric field E between two charged parallel plates. A positive test charge q is placed between the plates. The force on q is pointing to the [left ; right] and has the value [E ; qE ; E/q] . In order to do positive work on q it must be moved from plate [1 to 2 ; 2 to 1] . The test charge q will have [higher ; lower] potential energy when at Plate 1 rather than when at Plate 2. The work done in moving from one plate to the other will be [Ex ; qEx ; neither] . The potential energy difference ΔU between the two plates is [Ex ; qEx ; Ex/q ; neither] . If the charge q is free to move it will move toward plate [1 ; 2] , its potential energy will [increase ; decrease] and its kinetic energy will [increase ; decrease] . (Discussion continues in next frame.)

67
126

148

72

137

8
57
31
21

Go to 7-42.

7-42
(from 7-41)
The potential energy U increases as one moves to the [right ; left] and E is pointing to the [right ; left] , hence the direction of increasing U is the [same ; opposite] as that of E. Thus $\Delta U = $ [$qE\Delta x$; $-qE\Delta x$] . The electrostatic force would be
$F = \left[\Delta U/\Delta x \; ; \; -\Delta U/\Delta x \; ; \; -\frac{1}{q}\Delta U/\Delta x \; ; \; \frac{1}{q}\Delta U/\Delta x \; ; \; \text{neither} \right]$. In the study of electricity the potential V is defined as electric potential energy per unit charge. Hence
$V = \left[q\Delta U \; ; \; \frac{1}{q}\Delta U \; ; \; \text{neither} \right]$. Then $E = \left[\Delta V/\Delta x \; ; \; -\Delta V/\Delta x \; ; \; -q\Delta V/\Delta x \; ; \; -\frac{1}{q}\Delta V/\Delta x \; ; \; \text{neither} \right]$.

142
107
82
116

92

61
121

If incorrect go to 7-36 ; if correct go to 7-37.

7-43
(from 7-37)

The electric field inside the spherical shell is [Q/r^2 ; Q/R^2 ; neither] . **88**

The electric force on a test charge q inside the spherical shell is zero [*true* ; *false*]. **62**

The work done in moving a test charge q from the surface of the charged sphere to its center is the average force times the distance. This amount of work is zero [*true* ; *false*]. The potential difference between the surface and the center is zero **150**

[*true* ; *false*]. The potential at the center of a spherical shell is the same as the value **150**

at its surface [*true* ; *false*]. **129**

33

Go to 7-47.

7-44
(from 7-37)

The field between two parallel capacitor plates is $E = 4\pi Q/A$. If a dielectric slab of thickness d' is inserted between the plates, the field inside the dielectric will be $E' = \dfrac{4\pi Q}{\epsilon A}$, while the field outside will remain unchanged. The potential difference between the plates will now be $V =$

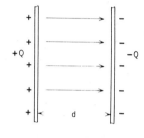

$\left[\; 4\pi Qd/A \;\;;\;\; 4\pi Qd/\epsilon A \;\;;\;\; \dfrac{4\pi Q}{A}\left(d+\dfrac{d'}{\epsilon}\right) \;\;;\;\; \dfrac{4\pi Q}{A}\left(d-d'+\dfrac{d'}{\epsilon}\right) \;\;;\;\; \text{neither} \;\right]$. **119**

The capacity will be $C = \left[\; \dfrac{A}{4\pi\left(d-d'+\dfrac{d'}{\epsilon}\right)} \;\;;\;\; 4\pi\left(d-d'+\dfrac{d'}{\epsilon}\right)/A \;\;;\;\; \text{neither} \;\right]$. **146**

If incorrect go to 7-46 ; if correct go to 7-27.

7-45
(from 7-21)

The electric field will be $E = Q/r^2$ only if the charged object Q is either a point charge or a uniformly charged sphere.

Go to 7-22.

7-46
(from 7-44)

The work done by a unit charge in moving the entire distance d will be the potential difference V. The work done by a unit charge in moving through the dielectric alone will be the force $E' = 4\pi Q/\epsilon A$ times the distance [d ; $d - d'$; d']. The additional distance to move between the dielectric and plates is [d ; $d - d'$; d']. In this region between dielectric and plates the force on a unit charge is [$4\pi Q/A$; $4\pi Q/\epsilon A$; neither] and the corresponding work done is $\left[\dfrac{4\pi Q}{A}(d - d') ; \dfrac{4\pi Q}{\epsilon A}(d - d') ; \right.$ neither $\left. \right]$. The total work done by the unit charge is the sum of the two above results or $\dfrac{4\pi Q d'}{A\epsilon} + \dfrac{4\pi Q}{A}(d - d')$ [*true* ; *false*]. This is also the value of V [*true* ; *false*].

64
97
90
131
99
160

Go to 7-47.

7-47
(from 7-43, 7-46, or 7-27)

Review Questions:

1. Fifteen lines of force cut through an area of 3 cm². What is the value of the component of E perpendicular to this area? $E = $ _____ dynes/statcoul.

2. In a certain direction the electric potential is increasing uniformly at the rate of 3 statvolts for each cm. What is the component of E in this direction?

   $E = $ _____ dynes/statcoul.

3. Does an electron in a Bohr orbit have a positive or negative potential energy? _____

   What is the relation between the potential and kinetic energy for such an electron?

   $KE = $ _____ U.

4. If at a given instant of time an electron in an elliptical orbit has $KE = 10^{-12}$ erg and $U = -3 \times 10^{-12}$ erg, its binding energy will be _____ erg.

5. The electric field between two concentric spheres does not depend on the charge on the outer sphere. True or false? _____

Chapter 7 Answer Sheet

Item Number	Answer
1	ball
2	left
3	less than
4	4
5	$4\pi\sigma$
6	$\dfrac{e^2}{r_c^2}$
7	$r + \dfrac{L}{2}$
8	qEx
9	$4\pi(Q_1 + Q_2 + Q_3)$
10	3.01×10^{14}
11	false
12	D
13	neither
14	negative
15	9×10^9
16	1.53×10^{-16}
17	positive
18	zero
19	$\dfrac{-e^2}{2r_c}$
20	less
21	increase
22	$\dfrac{\text{kg meter}^3}{\text{sec}^2 \text{ coul}^2}$
23	attracted
24	repelled
25	Q_2
26	$-W_0$
27	$\dfrac{e^2}{r^2}$
28	zero
29	hanging together
30	newtons
31	decrease
32	14.4×10^4
33	true
34	attracted
35	true
36	1
37	true
38	4
39	$4\pi(x + R_3)^2$
40	$\sqrt{GM_p/r}$
41	decrease
42	$Q_1 E$
43	$KE - e^2/r$
44	positive
45	right
46	negative
47	true

Item Number	Answer
48	true
49	$\sqrt{Fr^2}$
50	positive
51	$-\rho_2 E_1$
52	9×10^9
53	left
54	false
55	$\dfrac{-e^2}{r_0}$
56	more
57	2
58	coulombs
59	heavier
60	9×10^9
61	$\dfrac{1}{q}\Delta U$
62	true
63	$\dfrac{GM_p m}{r^2}$
64	d'
65	150
66	Q
67	right
68	$KE + U$
69	leaves
70	r_c
71	zero
72	higher
73	7.4×10^3
74	true
75	5
76	negative
77	than
78	4.8×10^{-10}
79	negative
80	larger
81	electric force
82	opposite
83	electrons
84	leave
85	leaves
86	$\dfrac{-e^2}{r_0}$
87	9×10^{14}
88	neither
89	$\dfrac{10}{10^2}$
90	$\dfrac{4\pi Q}{A}$
91	down
92	$\dfrac{-\Delta U}{\Delta x}$

Item Number	Answer	Item Number	Answer
93	$\dfrac{Q}{R}$	126	qE
94	the same as	127	come together
95	$\dfrac{(3 \times 10^9)^2}{100^2}$	128	$(Q_1 + Q_2 + Q_3)$
		129	true
96	$\text{gm}^{\frac{1}{2}} \text{cm}^{\frac{3}{2}} \text{sec}^{-1}$	130	$\dfrac{E}{\epsilon}$
97	$d - d'$	131	$\dfrac{4\pi Q}{A}(d - d')$
98	$\dfrac{Q_1 + Q_2 + Q_3}{(x + R_3)^2}$	132	false
		133	true
99	true	134	r_a to r_b
100	leaves	135	ball
101	true	136	$\dfrac{100}{20^2}$
102	$\dfrac{mv^2}{r}$	137	qEx
103	true	138	$\dfrac{2\rho_1 \rho_2}{R}$
104	out of	139	3×10^9
105	positive		
106	$\dfrac{Q}{(r + \frac{1}{2}L)^2}$	140	none
		141	repulsive
107	right	142	left
108	false	143	$4\pi(Q_1 + Q_2 + Q_3)$
109	positive	144	$\dfrac{1}{2}r_0$
110	$4\pi(Q_1 + Q_2)$		
111	the circumference by v	145	positive
112	$-\dfrac{1}{2}U$	146	$\dfrac{A}{4\pi\left(d - d' + \dfrac{d'}{\epsilon}\right)}$
113	3.01×10^{23}		
114	true		
115	$\dfrac{GM_p m}{r^2}$	147	$\dfrac{\sigma(\epsilon - 1)}{\epsilon}$
		148	2 to .1
116	$-qE\Delta x$		
117	neither	149	$\dfrac{Q}{(r - \frac{1}{2}L)^2} - \dfrac{Q}{(r + \frac{1}{2}L)^2}$
118	$\dfrac{1}{2}$	150	true
		151	true
119	$\dfrac{4\pi Q}{A}\left(d - d' + \dfrac{d'}{\epsilon}\right)$	152	q from ∞ to its present position
120	$4\pi(\sigma - \sigma')$	153	$\dfrac{2\rho_1}{R}$
121	$\dfrac{-\Delta V}{\Delta x}$	154	none
		155	positive
122	$\dfrac{Q}{R}$	156	$\dfrac{e^2}{2r_c}$
123	ball	157	$4\pi(\sigma - \sigma')$
124	true	158	positive
125	$\dfrac{e^2}{GM_p m}$	159	less
		160	true

Chapter 8
ELECTROMAGNETISM

Note: This chapter is the longest in this manual. It should take about twice the average amount of of time to complete.

8-1 Consider a tube of liquid. Suppose there are N positive ions per cm along the tube moving to the right with velocity v_1 (each ion has charge $+e$). Then there must be a current I flowing to the _____ with a value _____ .

A second tube has N negative ions per cm moving with velocity v_2 to the left. Then there must be a current flowing to the _____ with magnitude _____ .

A third tube containing a mixture of N positive and N negative ions per cm with respective velocities v_1 and v_2 carries a net current $I =$ [$Ne(v_1 + v_2)$; $Ne(v_1 - v_2)$; neither] flowing to the _____ .

40
50

16
25

55
81

Go to 8-2.

8-2 The electrostatic force between two charged wires varies as
(from 8-1) $\left[\dfrac{1}{r} \; ; \; \dfrac{1}{r^2} \; ; \; r \; ; \; r^2 \right]$.

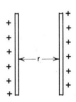

85

If incorrect go to 8-22 ; if correct go to 8-3.

8-3
(from 8-2 or 8-22)
The force between two parallel currents is proportional to $\left[\dfrac{1}{r}\ ;\ \dfrac{1}{r^2}\ ;\ r\ ;\ r^2\right]$ where r is the distance between the currents. The force between parallel currents (pointing in the same direction) is [attractive ; repulsive] .

77

47

Go to 8-4.

8-4
(from 8-3)
Suppose a new unit of current called the abampere, is defined by the equation $F = 2I_1 I_2 \ell_2 / r$ using CGS units for force and length; i.e., two wires 1 cm apart each carrying 1 abampere will exert an attractive force of _____ per unit length of the wire. Then 1 abampere would be _____ statcoul/sec.

10

4

If incorrect go to 8-12 ; if correct go to 8-5.

8-5
(from 8-4 or 8-12)
If we had defined a current element as $\dfrac{I\ell}{c}$ where I is in statcoul/sec, the magnetic force on the current element divided by the current element would be $\left[B\ ;\ \dfrac{B}{c}\ ;\ cB\ ;\ \text{none of these}\right]$.

86

Go to 8-6.

8-6
(from 8-5)
The force on a stationary test charge q will be $\left[B\ ;\ \dfrac{B}{c}\ ;\ E\ ;\ \text{none of these}\right]$.

95

If incorrect go to 8-15 ; if correct go to 8-7.

8-7
(from 8-6 or 8-15)

Consider a current I flowing into the page (it is perpendicular to the plane of the page). The magnetic field at point P_1 will be in the direction [a ; b ; c ; d ; in ; out] .

The magnetic field at point P_2 will be in the direction [e ; f ; g ; h ; in ; out] .

If r is the distance from P_1 to the current I, the value of the magnetic field at P_1 would be $B =$ _____ .

If there were a second current at P_1 pointing out of the page, the direction of magnetic force on this current would be along [a ; b ; c ; d ; in ; out] .

Go to 8-11.

69

65

36

42

8-8
(from 8-11)

The magnetic field at a distance of 1 meter from a current of 1 amp is
$$\left[\frac{2 \times 1}{1} \; ; \; \frac{2 \times 3 \times 10^9}{1} \; ; \; \frac{2 \times 3 \times 10^9}{100} \; ; \; \frac{.2}{100} \right] \text{ gauss.}$$

The force per meter length between two parallel currents of 1 amp each separated by 1 meter is _____ newtons per meter.

We see that the force between two unit currents in the MKS system is far from a unit force.

87

22

If incorrect go to 8-14 ; if correct go to 8-10.

8-9
(from 8-10)

The force on a current element has the value $F = \left[\left(\frac{I\ell}{c}\right) B \sin\theta \; ; \; \left(\frac{I\ell}{c}\right) B \; ; \; (I\ell)B \sin\theta \; ; \right.$ none of these $\left. \right]$ where θ is the angle between [ℓ and B ; ℓ and F ; F and B] .

A compass needle will point [perpendicular to B ; along B ; opposite to B ; none of these] .

58

91

73

Go to 8-13.

8-10
(from 8-8 or 8-14)

Right hand rule I gives the direction of magnetic [field produced by a current ; force on a current in a magnetic field ; neither] .   9

Right hand rule II gives the direction of the magnetic [field produced by a current ; field in a solenoid ; force on a current in a field ; none of these] .   44

The force on a current element in a magnetic field is always perpendicular to the current element [*true* ; *false*] .   21

The force on a current element in a magnetic field is always perpendicular to the field [*true* ; *false*] .   38

The current element is always perpendicular to the magnetic field in which it sits [*true* ; *false*] .   63

Go to 8-9.

8-11
(from 8-7)

Consider two currents I and I' flowing into the page. The force on current element $(I'\ell'/c)$ is _____ .   19

The force per unit current element is then _____ . This is the value of B produced by   84

the current I at the position of I' [*true* ; *false*] .   53

Go to 8-8.

8-12
(from 8-4)

In esu units the force between two identical currents is $F = \dfrac{2(I/c)^2 \ell}{r}$ [*true* ; *false*] .   79

If $F = 2$ dynes/cm of length and $r = 1$ cm, then $(I/c)^2 =$ _____ esu units, and I would   12

be _____ statcoul/sec.   27

Go to 8-5.

8-13
(from 8-9)

At the position of q the magnetic field B will point [toward the wire ; away from the wire ; in ; out ; up ; down] .

The positive charge q will experience a force [up ; down ; toward the wire ; away from the wire ; in ; out] .

The direction of the force is obtained from right hand rule [I ; II] . Using this rule, the thumb should be pointing [up ; down ; left ; right ; in ; out] and the fingers [along B ; perpendicular to B ; neither] .

The value of the force will be $F = \left[\dfrac{qvB}{c} \; ; \; qvB \; ; \; qB \; ; \; 0 \right]$.

It also has the value $\left[\dfrac{2qI}{c^2 r} \; ; \; \dfrac{2qI}{r} \; ; \; \dfrac{2qI}{cr} \; ; \; \text{none of these} \right]$.

If v is pointing toward the bottom of the page the direction of the force on q will be [up ; down ; in ; out ; left ; right] .

33

41

7
14
75

23

80

32

Go to 8-16

8-14
(from 8-8)

In esu units force is $\left(\dfrac{I\ell}{c} \right)$ times the value of B in gauss [true ; false] . For $I = 1$ amp and $B = \dfrac{.2}{100}$ gauss $F = \left[\dfrac{\ell}{10} \; ; \; \dfrac{\ell}{10c} \; ; \; \ell \right]$ times $\dfrac{.2}{100}$ dynes. For $\ell = 1$ meter the force is _____ dynes or _____ newtons.

61
31
46
57

Go to 8-10.

8-15
(from 8-6)

The force on a stationary test charge q will be $\left[E \; ; \; qE \; ; \; \dfrac{E}{q} \; ; \; \text{none of these} \right]$.

The force per unit charge q will be _____ .

The force per unit current element $(I\ell/c)$ will be [E ; B ; none of these] .

89

17

48

Go to 8-7.

8-16
(from 8-13)

A proton of charge *e* enters a region of uniform magnetic field. In region I where $B = 0$, the path of the proton will be [a straight line ; part of a circle] . In region II the path of the proton will be [a straight line ; part of a circle] .

Region I
$B = 0$

Region II
$B = 10^4$ gauss
out of page

11

1

As the proton enters region II there will be a force on it pointing [up ; down ; in ; out ; left ; right] .

43

The value of the force will be $\left[evB \; ; \; \dfrac{evB}{c} \; ; \; \text{neither} \right]$.

15

The value of the force will also be $\dfrac{mv^2}{R}$ where R is the radius of curvature of its circular path [*true* ; *false*] .

82

The value of R can be obtained from the relation $\dfrac{mv}{R} = \dfrac{eB}{c}$ [*true* ; *false*] .

52

While in region II the velocity of the proton will [increase ; decrease ; remain the same] .

59

While in region II the proton will [keep going around the same circle over and over again ; leave the region after traveling through a semicircle ; neither] .

92

The final velocity of the proton will be pointed [left ; right ; up ; down ; none of these] . In conclusion we see that charged particles in field-free space are in effect reflected from regions of magnetic field.

68

Go to 8-17.

8-17
(from 8-16)

Now assume q is moving into the page. The direction of the force will be [up ; down ; in ; out ; left ; right ; none of these] .

13

Go to 8-23

8-18
(from 8-23)

The force on a moving charge is $F = \dfrac{qvB}{c}$ only when v and B are perpendicular [*true* ; *false*] . If θ is the angle between v and B the general equation is $F = $ _____ .

71

37

Go to 8-19

8-19
(from 8-18)

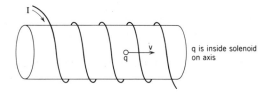

A positive charge q has a velocity v pointing along the axis of a solenoid. The direction of force on q is [up ; down ; left ; right ; in ; out ; none of these] . The value of the force on q is $\left[\dfrac{qvB}{c} \; ; \; qvB \; ; \; 0 \; ; \; \text{none of these} \right]$. The velocity is [perpendicular ; parallel] to B.

93

6
49

Go to 8-21.

8-20
(from 8-21)

Consider two wires with equal currents pointing into the page. The resultant magnetic field at point P will be $B = \left[\dfrac{2I}{cr} \; ; \; \dfrac{4I}{cr} \; ; \; 0 \; ; \; \text{none of these} \right]$ where r is the distance from P to each wire. Lines of B for the *resultant* magnetic field will look like [Fig. A ; B ; C] .

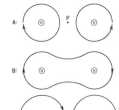

56

88

Go to 8-25.

8-21
(from 8-17)
Consider a current I pointing into the page. We draw a circle of radius R around it. Everywhere on this circle the value B is _____. The direction of B is pointing [along the circle ; perpendicular to the circle] .

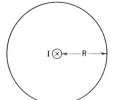

The average value of B along the circle is $\bar{B}_L$ = _____.

The total length of the circular closed path is L = _____.

Hence $\bar{B}_L \cdot L = \left[\; 4\pi I \; ; \; \dfrac{4\pi I}{c} \; ; \; \text{neither} \; \right]$.

The above relation happens to be true for a closed path of any shape. This relation is called Ampere's law.

Go to 8-20.

8-22
(from 8-2)
See frame 7-32 in Chapter 7 for the explanation.

Go to 8-3.

8-23
(from 8-17)
The direction of B is [in ; out ; up ; down ; left ; right]

The direction of B is [parallel ; perpendicular] to v.

The force on q is $\left[\; 0 \; ; \; \dfrac{qvB}{c} \; ; \; \text{neither} \; \right]$

Go to 8-18.

8-24
(from 8-26)
According to Ampere's law $\Sigma B_i L_i = \left[\; \dfrac{4\pi i}{c} \; ; \; \dfrac{4\pi i L_1}{c} \; ; \; 0 \; ; \; \dfrac{2\pi i}{c} \; ; \; \text{none of these} \; \right]$.

If incorrect go to 8-32 ; if correct go to 8-49.

8-25
(from 8-20)

Consider several parallel wires with equal currents into the page. The lines of force will look like this. The direction of B above the currents will be to the [left ; right] . The direction of B below the currents will be to the [left ; right] .

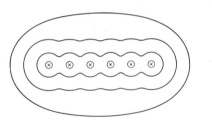

5

34

Go to 8-26.

8-26
(from 8-25)

Now consider a virtually infinite number of parallel currents pointing into the page.

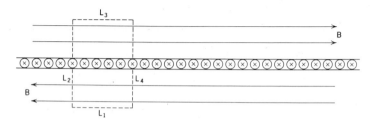

These currents make up a surface current of i statamps/cm where the surface is perpendicular to the page. The current flowing through the dashed rectangle will be

$$\left[\; i \; ; \; iL_1 \; ; \; \frac{i}{L_1} \; \right] \text{ statamps.}$$

60

The lines of B will be as shown. We wish to calculate the magnitude of B by applying [Ampere's ; Faraday's] law to the rectangular closed path. First we shall evaluate $B_1 L_1, B_2 L_2$, etc. where B_1 and B_2 are the components of B along L_1 and L_2 respectively. B_1 = [0 ; B] . B_2 = [0 ; B] .
Then $B_1 L_1 = B_3 L_3$ = [0 ; BL_1 ; $2BL_1$] and $B_2 L_2 = B_4 L_4$ = [0 ; BL_2 ; $2BL_2$] .

The complete sum around all 4 sides of the closed path is $\Sigma B_i L_i$ = [BL_1 ; $2BL_1$; 0 ; $4BL_1$; $2B(L_1 + L_2)$; none of these] .

30

24
90
78
62

8

Go to 8-24.

8-27
(from 8-49)

Now we shall consider two surface currents flowing in opposite directions.

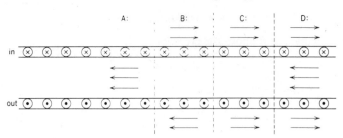

The resultant field lines will be as in Fig. [A ; B ; C ; D] and the net field between the two surfaces will be $B = \left[0 \; ; \; \dfrac{2\pi i}{c} \; ; \; \dfrac{4\pi i}{c} \; ; \; \text{none of these} \right]$.

64

76

If incorrect go to 8-30 ; if correct go to 8-28.

8-28
(from 8-27 or 8-30)

If the two conducting planes are of finite length, the lines of B will be shown. The magnitude of B between the planes will still be $B = \left[\dfrac{2\pi i}{c} \; ; \; \dfrac{4\pi i}{c} \; ; \; \dfrac{2i}{cr} \; ; \; 0 \right]$

35

except near the ends. Note that the above situation is quite similar to that of the side view of a solenoid of rectangular cross section.

Let us now consider the above solenoid of n turns per cm where each turn carries a current I. Then the corresponding surface current would be $i =$

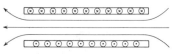

$\left[nI \; ; \; \dfrac{I}{n} \; ; \; I \; ; \; \text{none of these} \right]$ statamps/cm.

20

The magnetic field inside would then be $B = \dfrac{4\pi (nI)}{c}$ [*true* ; *false*] . Note that we have arrived at the formula for the field inside a solenoid.

28

Go to 8-29.

8-29
(from 8-28)

We wish to calculate the forces on a single rectangular loop of current when in an external magnetic field B. Consider side a_1 where the current is flowing into the page.

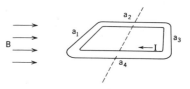

The force on a_1 will be $F_1 =$
$\left[\dfrac{IB}{c} \ ; \ 0 \ ; \ \dfrac{Ia_1 B}{c} \ ; \ \text{none of these} \right]$ in the [up ; down ; left ; right] direction.

The force on a_3 will have the same magnitude as F_1 [*true* ; *false*] . The force on a_3 will have the [same ; opposite] direction as F_1. The two forces together will tend to rotate the loop [clockwise ; counterclockwise] about the dotted line axis. The force on a_2 will be [in ; out ; up ; down ; left ; right ; zero] . The forces on a_2 and a_4 [will ; will not] tend to rotate the loop.

74
66
83
26
29
72
70

Go to 8-31.

8-30
(from 8-27)

In order to determine the resultant lines of B, let us first consider the lines produced by each surface current separately. In this figure the dashed lines are due to the [upper ; lower] surface current only. The heavy lines are lines of B produced by the [upper ; lower] surface current only. In regions I and III the sum of the heavy and dashed lines will [add ; cancel] . The net field in regions I and III will be $B = \left[\dfrac{2\pi i}{c} \ ; \ \dfrac{4\pi i}{c} \ ; \ 0 \right]$. The net field in region II will be $B = \left[\dfrac{2\pi i}{c} \ ; \ \dfrac{4\pi i}{c} \ ; \ 0 \ ; \ \text{none of these} \right]$.

67
128
96
177
110

Go to 8-28

8-31 (from 8-29) Torque is defined as distance from the axis times the force. The torque due to F_1 is then $T_1 = \left[F_1 \ ; \ F_1 a_1 \ ; \ \frac{1}{2} F_1 a_1 \ ; \ F_1 a_1 \ ; \right.$ 136

$\left. \frac{1}{2} F_1 a_2 \right]$. The total torque due to F_1 and F_3 is $[F_1 a_1 \ ; \ F_1 a_2 \ ; \ 2 F_1 a_2]$. 153

This has the value $T = IBa_1 a_2/c$ [true ; false]. 184

The torque also has the value $T = IBA/c$ where $A = a_1 a_2$ is the area of the coil

[true ; false]. It can be shown that this formula holds for current loops of any shape 105

[true ; false]. 114

Go to 8-33.

8-32 (from 8-24) The total current enclosed by the rectangle in frame 8-26 is $I = [\ i \ ; \ iL_1 \ ; \ 0 \ ; \text{ none of}$ 161

these] . 173

The right hand side of Ampere's law should be $\left[\dfrac{4\pi i}{c} \ ; \ \dfrac{4\pi I}{c} \right]$.

Go to 8-49.

8-33 (from 8-31) Suppose there are N current loops of area A each mounted together as shown. The total torque on the entire system would be $T = \left[\dfrac{IBA}{c} \ ; \ \dfrac{NIBA}{c} \ ; \text{ neither} \right]$. 124

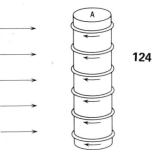

Go to 8-38.

8-34 (from 8-39) The lines of force inside this solenoid point [left ; right] . 178

The equivalent positive magnetic pole is at the [top ; 149

bottom ; left end ; right end] .

If an external magnetic field is running from the bottom to the 101

top of this page the torque on the solenoid will tend to rotate it

[clockwise ; counterclockwise] .

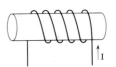

Go to 8-35

8-35 (from 8-34) This is the bar magnet that is equivalent to the solenoid of the previous frame. The force on the left end of this bar magnet has a value $\left[\ mB\ ;\ \dfrac{B}{m}\ \right]$ and is pointing [along B ; opposed to B ; neither] .

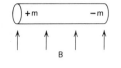

100
142

The force on the right end is pointing [along B ; opposed to B ; neither] . 152

These two forces tend to rotate the bar magnet [clockwise ; counterclockwise] and line it up [parallel ; antiparallel] to B .

109
121

Go to 8-36.

8-36 (from 8-35) This single loop is connected to a battery through the fixed brushes a and b. In the position shown there will be a torque tending to rotate the loop [clockwise ; counterclockwise] about the dashed line axis. After the loop has rotated 180° there will be a torque tending to rotate the loop [clockwise ; counterclockwise] . This frame illustrates the principle of the electric [motor ; generator] .

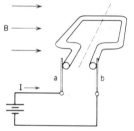

150

111
107

Go to 8-41

8-37 (from 8-41) Emf has the units [dynes/statcoul ; erg/statcoul ; dynes ; ergs ; ergs/sec] . 123

Emf has the same units as electric [field ; potential ; potential energy ; none of these] . 138

Go to 8-40.

8-38 (from 8-33) A solenoid has N turns over a length L. Let i be the current per cm of length. Then $i = \left[\ NI\ ;\ \dfrac{NI}{L}\ ;\ \dfrac{I}{L}\ \right]$.

The torque on the solenoid is $T = \left[\ \dfrac{iBAL}{c}\ ;\ \dfrac{iBA}{Lc}\ ;\ \dfrac{iBA}{c}\ ;\ \text{none of these}\ \right]$.

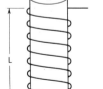

166
171

Go to 8-39.

8-39
(from 8-38)

We know that a solenoid can be represented by an equivalent bar magnet of pole strength m. The force on each pole is $\left[mB \; ; \; \dfrac{mB}{c} \; ; \; 2mB \; ; \text{ none of these} \right]$. The torque on this equivalent bar magnet is

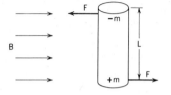

$T = \left[FL \; ; \; 2FL \; ; \; \dfrac{1}{2} FL \right]$.

This has the value $\left[mBL \; ; \; 2mBL \; ; \; \dfrac{1}{2} mBL \right]$.

146
140
99

According to the preceding frame this same torque in terms of the current i has the value $iBAL/c$ [*true* ; *false*] .

113

If we equate the two previous answers we obtain the relation $m = $ _____ , in terms of the solenoid current per cm length along the solenoid. Conversely, if we have a bar magnet of pole strength m, we can use the above formula to calculate i which represents the net atomic current that must be perpetually circulating around the outer surface of the ferromagnetic material.

119

Go to 8-34.

8-40
(from 8-37)

Suppose the current for a solenoid is turned on by throwing a switch. If the current is increasing from zero, the number of lines of B in the solenoid will be increasing [*true* ; *false*]. If there is an increase in the N_B enclosed by the turns of the solenoid there will be an induced emf $= \dfrac{1}{c} \dfrac{\Delta N_B}{\Delta t}$ for each turn of the solenoid

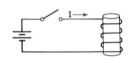

167

[*true* ; *false*] . The total induced emf will be $\left[\dfrac{N}{c} \dfrac{\Delta N_B}{\Delta t} \; ; \; \dfrac{1}{c} \dfrac{\Delta N_B}{\Delta t} \; ; \; \dfrac{N^2}{c} \dfrac{\Delta N_B}{\Delta t} \right]$

143
108

where N is the total number of turns of the solenoid. This induced emf is called the back emf V'. V' is proportional to $\left[I \; ; \; N_B \; ; \; \dfrac{\Delta I}{\Delta t} \; ; \text{ none of these} \right]$.

127

If incorrect go to 8-45 ; if correct go to 8-42.

8-41
(from 8-36)

Consider a rotating loop as shown. The force on a conduction electron at point P will have the value $\left[\, 0 \,;\, \dfrac{Bv}{c} \,;\, \dfrac{evB}{c} \,;\, \text{none of these} \,\right]$ and be pointing [in ; out ; up ; down ; none of these] . The force on the electron will tend to move it toward [a ; b] . The electric potential at wire a. will be [more positive than ; more negative than ; the same as] at b. When a. is passing through the 90° position (after $\dfrac{1}{4}$ turn) the potential at a. will be [more positive ; more negative ; the same as] at b. This illustrates the principle of the [ac ; dc] [generator ; motor] .

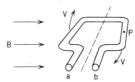

151

188

175
120

106

115
191

Go to 8-37.

8-42
(from 8-40)

The proportionality constant between the back emf V' and $\Delta I/\Delta t$ is called the self-inductance L. $V' = L\left(\dfrac{\Delta I}{\Delta t}\right)$. Now we shall derive a formula for the self-inductance of a solenoid in terms of its size and number of turns N.

According to Faraday's law $V' = \left[\, \dfrac{N}{c}\dfrac{\Delta N_B}{\Delta t} \,;\, \dfrac{1}{c}\dfrac{\Delta N_B}{\Delta t} \,;\, \dfrac{N}{c}\dfrac{\Delta 1}{\Delta t} \,\right]$.

N_B has the value $\left[\, B \cdot A \,;\, \dfrac{B}{A} \,\right]$.

B has the value $\left[\, \dfrac{4\pi I}{cN} \,;\, \dfrac{4\pi I}{c}\left(\dfrac{N}{\ell}\right) \,;\, \dfrac{4\pi IN}{c} \,\right]$. Multiplying by A gives $N_B = \dfrac{4\pi NAI}{c\ell}$

[true ; false] . Hence $\dfrac{\Delta N_B}{\Delta t} =$ _____ .

The back emf V' is N/c times this or $V' = \left[\, \dfrac{4\pi N^2 A}{c^2 \ell} \,;\, \dfrac{4\pi A}{\ell} \,;\, \dfrac{4\pi N^2}{c^2 \ell^2} \,\right]$ times $(\Delta I/\Delta t)$.

Hence $L =$ _____ .

162
145

187

159
154

144

148

Go to 8-44.

8-43
(from 8-45)

Faraday's law states that a changing [magnetic ; electric] field induces an emf.
Faraday's law also states that changing [magnetic ; electric] field induces a [magnetic ; electric] field.

Emf can be expressed in terms of the induced electric field [true ; false] .

Emf = $\left[\, \Sigma E_i \cdot L_i \,;\, E \,;\, \dfrac{E_L L}{\Sigma L_i} \,\right]$ around a closed path.

104
181
125
157

184

133

Go to 8-46.

8-44
(from 8-42)
A 6 volt battery is suddenly connected across a 1000 turn solenoid that produces 10^5 lines of force for every ampere passing through it. $N_B = 10^5 I$ lines/amp. We will now see how I must increase with time after the switch is closed. Let V' stand for the magnitude of the back emf and V_0 for the battery voltage. Then the net voltage tending to drive current around the circuit is [$V_0 - V'$; $V_0 + V'$] .   97

Suppose V' is less than V_0 and that the wire and battery resistances are zero; then the current I would be [zero ; infinite] . In order for I to be finite, we conclude that V' must be equal to V_0 (assuming zero resistance). If $V' = V_0$, V' is constant and $\Delta I/\Delta t$ must also be constant [*true* ; *false*] . In this problem we wish to calculate $\Delta I/\Delta t$ or the rate of increase of the current I.   130
    135

$$V' = \frac{N}{c}\frac{\Delta N_B}{\Delta t} = \frac{N}{c}\left(10^5 \frac{\Delta I}{\Delta t}\right)$$ for I in [statamps ; amperes] and V' in [volts ; statvolts] . Solving for $\Delta I/\Delta t$ we obtain $\Delta I/\Delta t = cV'/10^5 N$ where $V' =$ 6 volts = _____ statvolts.   103
    116
    185

$\frac{\Delta I}{\Delta t} =$ _____ amps/sec.   179

Three seconds after the switch is thrown the current will be _____ amps.   190

What is the self-inductance of the above solenoid in MKS units? In the MKS system we have the same equation $V' = L(\Delta I/\Delta t)$, but now V' is in volts and I in amperes. The unit of L has a special name, the henry. $L =$ _____ henrys.   205

Go to 8-45.

8-45
(from 8-40 or 8-44)

Consider a conducting rod carrying a current *I*. We know that the magnetic field outside the conducting rod at distance *r* from the axis is B _____. 169

But what is the value of *B* inside the rod? 98

_____ law is usually used to calculate magnetic fields. For the closed path take a circle of radius *r* inside the rod as shown. If B_r is the value of *B* at a distance *r* from the axis $\Sigma B_i L_i =$ [$B_r 2\pi r$; $B_r \pi r$; $2B_r r$; none of these]. This must be equal to $4\pi I_r/c$ where I_r is the [total current *I* ; current at *r* only ; current enclosed by the circle of radius *r*] . 112 / 117

Assuming the current is uniform over the cross section of the rod, $I_r =$

$$\left[I \; ; \; \frac{Ir}{R} \; ; \; \frac{Ir^2}{R^2} \; ; \; \text{none of these} \right].$$ Hence $\Sigma B_i L_i = \dfrac{4\pi}{c} \dfrac{r^2}{R^2} I.$ 155

Substituting:

$$(B_r \cdot 2\pi r) = \frac{4\pi r^2 I}{cR^2}$$

And

$$B_r = \frac{2Ir}{cR^2} \, .$$

Go to 8-43.

8-46
(from 8-43)

A changing magnetic field is produced when an electromagnet is turned on. Assume for this magnet that $\Delta B/\Delta t = 10^4$ gauss/sec.

While *B* is increasing there [will ; will not] be an electric field between the poles of the electromagnet. 201

After the electromagnet is fully turned on (*B* is constant), there [will ; will not] be an electric field between the poles of the magnet. 176

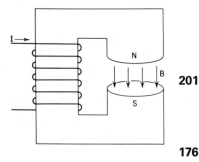

Go to 8-47.

8-47
(from 8-46)
Let us calculate the value of E at a distance of r from the center of the circular pole face. Consider a closed path as shown of length $L = $ _____ .

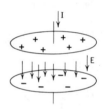

The emf around the path will be $\dfrac{1}{c}\dfrac{\Delta N_B}{\Delta t}$ where $N_B =$ [$B\pi r^2$; $2B\pi r^2$; $B\pi R^2$; $2B\pi R^2$] . Also the emf has the value $\left[2\pi r E \; ; \; rE \; ; \; \dfrac{2\pi rE}{L} \right]$.

Equating the two above expressions for emf gives $E = $ _____ .

For $r = 10$ cm and $\Delta B/\Delta t = 10^4$ gauss/sec we have $E = $ _____ dynes/statcoul.

The direction of E would be [parallel ; perpendicular] to B.

118
156
189
164
182
126

Go to 8-48

8-48
(from 8-47)
Suppose we turned on an electric field between two circular metal plates by running a current or charges into one plate and out of the other. While E was increasing there [would ; would not] be a magnetic field between the metal plates. This field can be calculated using [Faraday's law ; Maxwell's modification of Ampere's law] .

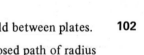

If E were constant, there [would ; would not] be a magnetic field between plates. As in the previous frame let us now calculate B by taking a circular closed path of radius r between the two plates where r is less than the radius of the plates. The equation to use is $\left[\bar{E}_L \cdot L = \dfrac{1}{c}\left(\dfrac{\Delta N_B}{\Delta t}\right) \; ; \; \bar{B}_L \cdot L = \dfrac{4\pi I}{c} + \dfrac{1}{c}\left(\dfrac{\Delta N_E}{\Delta t}\right) \right]$

The current I flowing through the closed path is $\left[0 \; ; \; E \cdot \pi r^2 \; ; \; \pi r^2\left(\dfrac{\Delta E}{\Delta t}\right) \; ; \; \text{none of these} \right]$.

The left hand side has the value [$B\pi r^2$; $2\pi Br$; neither] . The right hand side has the value $\left[\dfrac{\pi r^2}{c}(\Delta B/\Delta t) \; ; \; \dfrac{\pi r^2}{c}(\Delta E/\Delta t) \; ; \; \text{neither} \right]$.

The solution is $B = \left[\dfrac{\pi r}{c}(\Delta E/\Delta t) \; ; \; \dfrac{\pi r}{2c}(\Delta E/\Delta t) \; ; \; \dfrac{r}{2c}(\Delta E/\Delta t) \right]$.

For $r = 10$ cm and $(\Delta E/\Delta t) = 10^4$ esu/sec we have $B = \dfrac{10^5}{6 \times 10^{10}}$ [gauss ; MKS units] .

The direction of B is [parallel ; perpendicular] to E.

131
168
102
180
186
160
141
192
172
147

Go to 8-50.

8-49
(from 8-24 or 8-32)

We have the relation (from frame 8-24) $\Sigma B_i L_i = \dfrac{4\pi i}{c} L_1$.

The left hand side was evaluated to be $2BL_1$ (in frame 8-26).

Equating these two results gives $B =$ _____ . 122

This is the magnetic field produced by a [solenoid ; surface current] . 134

Go to 8-27.

8-50
(from 8-48)

In these next two frames we will show that Maxwell's equations require that an electromagnetic wave must travel through empty space with the speed of light.

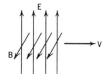

Assume a region of crossed electric and magnetic fields is traveling to the right in free space with a velocity v. The figure shows the fields when they have traveled a distance x. Beyond x the fields are zero. The velocity of the wave will be $v = \dfrac{\Delta x}{\Delta t}$. We wish to calculate v and the relation between E and B.

First take a vertical rectangle of sides L_1, L_2, L_3, and L_4 as a closed path. Lines of [E ; B] are enclosed by this path. The appropriate equation to use would then be [Ampere's ; Faraday's] Law. The emf around this closed path is [0 ; EL_3 ; BL_3 ; $E(L_2 + L_3 + L_4)$; none of these] . 163 137 158

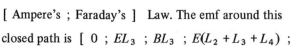

This must be equal to $\left[\dfrac{1}{c}(\Delta N_B/\Delta t) \; ; \; \dfrac{1}{c}(\Delta N_E/\Delta t) \; ; \; \text{neither} \right]$. 174

The number of lines of B cutting through the rectangle is $N_B =$ [$BL_1 L_2$; $BL_3 x$; neither] . 209

Then $\dfrac{\Delta N_B}{\Delta t} = \dfrac{BL_3 \Delta x}{\Delta t}$ [true ; false], where $\dfrac{\Delta x}{\Delta t} =$ [v ; B ; E ; none of these] . 132 170

Faraday's law says $\Sigma E_i L_i = \dfrac{1}{c}\left(\dfrac{\Delta N_B}{\Delta t}\right)$ [true ; false]. 139

The right hand side has the value $\left[\dfrac{BvL_3}{c} \; ; \; BvL_3 \; ; \; BvL_2 L_3 \right]$. 129

The left hand side has the value _____ . 165

Equating the two above results gives $\left[E = Bv \; ; \; E = \dfrac{Bv}{c} \; ; \; B = \dfrac{Ev}{c} \; ; \; \text{none of these} \right]$. 183

Go to 8-51.

8-51
(from 8-50)
Now take the same rectangle, but rotated 90° about L_2 so that it is now perpendicular to the lines of E. Lines of [E ; B] will be enclosed by this [horizontal ; vertical] rectangle. Hence the appropriate equation to use is [Ampere's law (as modified by Maxwell) ; Faraday's law] .

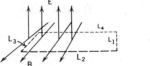

193
202
214

The quantity $\Sigma B_i \cdot L_i$ has the value [BL_3 ; $B(L_3 + 2L_2)$; neither] . 197

The quantity $\dfrac{1}{c}\left(\dfrac{\Delta N_E}{\Delta t}\right)$ has the value $\dfrac{1}{c}\dfrac{EL_3\Delta x}{\Delta t}$ which is $\dfrac{EvL_3}{c}$ [true ; false]. 199

For $I = 0$ the modified Ampere's law is $\Sigma B_i \cdot L_i = \dfrac{1}{c}\dfrac{\Delta N_E}{\Delta t}$ [true ; false]. 211

The left hand side has the value BL_3 [true ; false]. 196

The right hand side has the value $\dfrac{EvL_3}{c}$ [true ; false]. 206

Equating the two above results gives $B = \dfrac{Ev}{c}$ which is [the same as ; different from] the final result of the previous frame. If we multiply these two final results by each other we obtain $\left[EB = \dfrac{EBv^2}{c^2} \ ; \ EB = BE \right]$. 213

194

Hence $v = c$ and $E = B$ must be true for a traveling electromagnetic wave.

Go to 8-52.

8-52   Review Questions:

(from 8-51)

1. What are the units of emf? _____   **195**

2. Which is emf ($\Sigma E_i \cdot \ell_i$; ΣB_i ; ℓ_i ; or $\Sigma F_i \cdot \ell_i$)? _____   **212**

3. What are the esu units of current? _____   **203**

4. Write Ampere's law. _____   **200**

5. Write Ampere's law as modified by Maxwell. _____   **207**

6. Write Faraday's law. _____   **198**

7. Which of these laws is used to calculate magnetic field strengths due to steady currents? _____   **210**

8. Which rule is used to determine the direction of the magnetic field produced by steady currents? _____   **204**

9. A magnet having a pole strength of $m = 3$ cgs units is sitting in a magnetic field B. If the force on each magnetic pole is 18 dynes, what is B? $B = $ _____ gauss.   **208**

Chapter 8 Answer Sheet

Item Number	Answer	Item Number	Answer	Item Number	Answer
1	part of a circle	45	$\dfrac{2I}{cR}$	89	qE
2	$\dfrac{2I}{cr}$	46	.02	90	zero
3	parallel	47	attractive	91	l and B
4	3×10^{10}	48	B	92	leave the region
5	right	49	parallel	93	neither
6	zero	50	$Nv_1 e$	94	in
7	II	51	$\dfrac{4\pi i L_1}{c}$	95	neither
8	$2BL_1$			96	cancel
9	field produced by a current	52	true	97	$V_0 - V'$
10	2 dynes	53	true	98	Ampere's
11	straight line	54	$2\pi R$	99	mBL
12	1	55	$Ne(v_1 + v_2)$	100	mB
13	neither	56	zero	101	clockwise
14	left	57	2×10^{-7}	102	would not
15	$\dfrac{evB}{c}$	58	$\dfrac{(Il)}{c} B \sin\theta$	103	amperes
				104	magnetic
16	right	59	remain the same	105	true
17	E	60	iL_1	106	the same as
18	zero	61	true	107	motor
19	$\dfrac{2I(I'l')}{c^2 r}$	62	zero	108	$\dfrac{N}{c} \dfrac{\Delta N_B}{\Delta t}$
		63	false		
20	nI	64	A	109	clockwise
21	true	65	e	110	$\dfrac{4\pi i}{c}$
22	2×10^{-7}	66	down	111	counterclockwise
23	$\dfrac{qvB}{c}$	67	lower	112	$B_r 2\pi r$
24	B	68	left	113	true
25	Nev_2	69	d	114	true
26	opposite	70	will not	115	ac
27	3×10^{10}	71	true	116	statvolts
28	true	72	zero	117	enclosed current
29	counterclockwise	73	along B	118	$2\pi r$
30	Ampere's	74	$\dfrac{Ia_1 B}{c}$	119	$\dfrac{iA}{c}$
31	$\dfrac{l}{10}$	75	along B	120	more negative than
32	right	76	$\dfrac{4\pi i}{c}$	121	parallel
33	in	77	$\dfrac{1}{r}$	122	$\dfrac{2\pi i}{c}$
34	left	78	BL_1	123	erg/statcoul
35	$\dfrac{4\pi i}{c}$	79	true	124	$\dfrac{NIBA}{c}$
36	$\dfrac{2I}{cr}$	80	none	125	electric
		81	right	126	perpendicular
37	$\dfrac{qv}{c} B \sin\theta$	82	true	127	$\dfrac{\Delta I}{\Delta t}$
		83	true		
38	true	84	$\dfrac{2I}{cr}$	128	upper
39	along the circle	85	$\dfrac{1}{r}$	129	$\dfrac{BvL_3}{c}$
40	right	86	B	130	infinite
41	down	87	$\dfrac{.2}{100}$	131	would
42	c			132	true
43	down	88	B	133	$\Sigma E_i \cdot L_i$
44	force on a current in a field			134	surface current

Item Number	Answer	Item Number	Answer	Item Number	Answer
135	true	163	B	188	in
136	$\frac{1}{2}F_1a_2$	164	$\frac{r}{2c}\left(\frac{\Delta B}{\Delta t}\right)$	189	$2\pi rE$
137	Faraday's	165	EL_3	190	18
138	potential	166	$\frac{NI}{L}$	191	generator
139	true			192	$\frac{r}{2c}\left(\frac{\Delta E}{\Delta t}\right)$
140	FL	167	true	193	E
141	$\frac{\pi r^2}{c}\left(\frac{\Delta E}{\Delta t}\right)$	168	Maxwell's modification	194	$EB = \frac{EBv^2}{c^2}$
142	along B	169	$\frac{2I}{cr}$	195	statvolts
143	true	170	v	196	true
144	$\frac{4\pi N^2 A}{c^2 l}$	171	$\frac{iBAL}{c}$	197	BL_3
145	$B \cdot A$	172	gauss	198	$\Sigma E_i \cdot l_i = \frac{1}{c}\left(\frac{\Delta N_B}{\Delta t}\right)$
146	mB	173	$\frac{4\pi I}{c}$	199	true
147	perpendicular			200	$\Sigma B_i \cdot l_i = \frac{4\pi I}{c}$
148	$\frac{4\pi N^2 A}{c^2 l}$	174	$\frac{1}{c}\left(\frac{\Delta N_B}{\Delta t}\right)$	201	will
149	left end	175	a	202	horizontal
150	counterclockwise	176	will not	203	statamps
151	$\frac{evB}{c}$	177	zero	204	Right hand rule I
		178	left	205	1
152	opposed to B	179	6	206	true
153	F_1a_2	180	$B_L \cdot L = \frac{4\pi I}{c} + \frac{1}{c}\left(\frac{\Delta N_E}{\Delta t}\right)$	207	$\Sigma B_i \cdot l_i = \frac{4\pi I}{c} + \frac{1}{c}\left(\frac{\Delta N_E}{\Delta t}\right)$
154	$\frac{4\pi NA}{cl}\frac{\Delta I}{\Delta t}$	181	magnetic	208	6
155	$\frac{Ir^2}{R^2}$	182	$\frac{1}{6} \times 10^{-5}$	209	$BL_3 x$
156	$B\pi r^2$	183	$E = \frac{Bv}{c}$	210	Ampere's law
157	true			211	true
158	EL_3	184	true	212	$\Sigma E_i \cdot l_i$
159	true	185	.02	213	different from
160	$2\pi Br$	186	zero	214	Ampere's law, modified
161	iL_1	187	$\frac{4\pi I}{c}\left(\frac{N}{l}\right)$	215	$\frac{4\pi I}{c}$
162	$\frac{N}{c}\frac{\Delta N_B}{\Delta t}$				

Chapter 9
ELECTRICAL APPLICATIONS

9-1   (1 volt) × (1 coulomb) = [10^7 joule ; 1 joule ; 10^5 erg ; 1 watt] .   **96**

(1 statvolt) × (1 statcoul) = [1 watt ; 1 joule ; 1 erg ; 1 ohm] .   **73**

1 statamp = $\left[\dfrac{1}{3} \times 10^{10} \;;\; \dfrac{10}{3 \times 10^{10}} \;;\; 3 \times 10^9 \right]$ amperes.   **107**

1 statvolt = $\left[3 \times 10^9 \;;\; 3 \times 10^{10} \;;\; 3 \times 10^2 \;;\; \dfrac{1}{300} \right]$ volts.   **4**

Go to 9-4.

9-2   The resistance of a metal bar is [directly ; inversely] proportional to its length.   **29**

(from 9-4)   If the cross sectional area is doubled, the resistance will be [halved ; doubled ; neither] .   **38**

Go to 9-6.

9-3
(from 9-11)

R_t = [2 ; 8 ; neither] ohms.   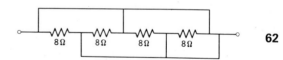   **62**

If incorrect go to 9-13 ; if correct go to 9-12.

9-4
(from 9-1)

If an electric field E exists inside a metal, a current will flow in the [*same* ; *opposite*] direction as the electric lines of force. The electrical conductivity of a metal is defined as the ratio of the current per unit area divided by the electric field. $\sigma = \dfrac{i}{E}$ where σ is the conductivity and i is the current per unit area. The value of σ will depend on E [*true* ; *false*] .

The value of σ will depend on the temperature [*true* ; *false*] . For a given metal, the value of σ will depend on the particular shape [*true* ; *false*] .

We wish to find the resistance of a rectangular bar of length L and cross sectional area A in terms of its conductivity σ. If a voltage V is applied across the length L, the electric field inside the metal bar will be $\left[0 \; ; \; V \; ; \; \dfrac{V}{L} \; ; \; VL \right]$. The resistance is defined as $R = \left[\dfrac{V}{I} \; ; \; \dfrac{I}{V} \; ; \; IV \right]$. Hence $R = \dfrac{EL}{I}$ [*true* ; *false*] . Since i is the current per unit area, the total current $I = \left[i \; ; \; iA \; ; \; \dfrac{i}{A} \right]$. Substituting this into the above equation for R gives $R =$ _____ or $R = \left[\dfrac{\sigma L}{A} \; ; \; \dfrac{L}{\sigma A} \; ; \; \sigma AL \; ; \; \dfrac{AL}{\sigma} \; ; \; \text{none of these} \right]$.

5

95

106

11

14

75
88

47

64
80

Go to 9-2.

9-5
(from 9-8)

The total resistance is $R_t = \left[R_1 + R_2 + R_3 \; ; \; \dfrac{1}{\dfrac{1}{R_1} + \dfrac{1}{R_2} + \dfrac{1}{R_3}} \; ; \; \text{neither} \right]$.

17

$\dfrac{1}{R_1} + \dfrac{1}{R_2} + \dfrac{1}{R_3} = \left[\dfrac{R_1 + R_2 + R_3}{R_1 R_2 R_3} \; ; \; \dfrac{R_3 R_2 + R_1 R_3 + R_2 R_1}{R_1 R_2 R_3} \; ; \; \dfrac{1}{R_1 + R_2 + R_3} \; ; \; \text{none of these} \right]$.

56

Go to 9-9.

9-6
(from 9-2)
Consider a metal wire of cross sectional area A. Let η be the number of conduction electrons per cm³ of the metal. Then the moving charge per unit length along the wire will be $\rho = \left[\eta A \; ; \; \eta e \; ; \; \eta A e \; ; \; \dfrac{\eta e}{A} \; ; \; \text{none of these} \right]$.

86

If the average drift velocity of this charge is $\bar{v}_d$, the number of conduction electrons moving by in one second will be _____ . The current I will be _____ . The current density (current per unit area) will be $i = \left[\eta e A \bar{v}_d \; ; \; \dfrac{\eta e v_d}{A} \; ; \; \eta e \bar{v}_d \; ; \; \text{none of these} \right]$.

34
102

85

Go to 9-7.

9-7
(from 9-6)
Newton's second law gives $m \dfrac{\Delta u}{\Delta t} = eE$. Using $\Delta t = \dfrac{L}{u}$, we obtain $\Delta u = \dfrac{eLE}{mu}$, which was derived where L is the mean free path between collisions for the conduction electrons. Let η be the number of conduction electrons per cm³ of the metal. The average drift velocity will be $\bar{v}_d = \left[u \; ; \; \dfrac{u}{2} \; ; \; \Delta u \; ; \; \dfrac{\Delta u}{2} \right]$. Then i, the current/cm², will be $i = \left[\eta \left(\dfrac{\Delta u}{2}\right) \; ; \; \eta L \left(\dfrac{\Delta u}{2}\right) \; ; \; \eta e \left(\dfrac{\Delta u}{2}\right) \; ; \; \text{none of these} \right]$ with the result that $i = \dfrac{\eta e^2 LE}{2mu}$

90

13

[*true* ; *false*] . The conductivity σ which is $\left[\dfrac{E}{i} \; ; \; \dfrac{i}{E} \; ; \; iE \right]$ will then be $\left[\dfrac{\eta eL}{2mu} \; ; \; \dfrac{\eta e^2 L}{2mu} \; ; \; \dfrac{\eta eLE}{2mu} \; ; \; \text{none of these} \right]$. Using the theory of conduction electrons in a metal, we have derived a formula for the conductivity of any metal in terms of the velocity, density, and mean free path of the metal's conduction electrons. The resistance of a metal is [*directly* ; *inversely*] proportional to its conductivity.

27
70

45

21

Go to 9-8.

9-8
(from 9-7)

The total resistance is $R_t = \left[R_1 + R_2 \; ; \; \frac{1}{2}(R_1 + R_2) \; ; \; \frac{1}{R_1} + \frac{1}{R_2} \; ; \; \frac{R_1 R_2}{R_1 + R_2} \; ; \;$ none of these $\right]$.

99

Go to 9-5.

9-9
(from 9-5)

$I_2 = [\; I_1 + I_3 \; ; \; I_1 - I_3 \; ; \; I_3 - I_1 \;]$.

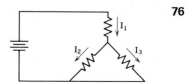

76

Go to 9-10

9-10
(from 9-9)

The three resistors are connected in [series ; parallel ; neither] .

If the wire should break at point A the total resistance will be _____ .

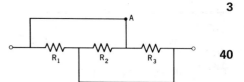

3

40

Go to 9-11.

9-11
(from 9-10)

$R_t = [\; 2 \; ; \; 10 \; ; \; 50 \; ;$ none of these $]$ ohms.

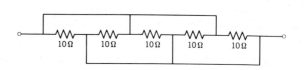

35

Go to 9-3.

9-12
(from 9-3 or 9-13)

The resistance of this parallel circuit is _____ ohms. 32

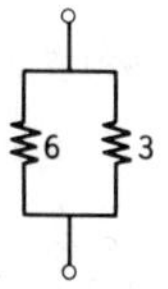

The total resistance of this circuit is _____ ohms. 8

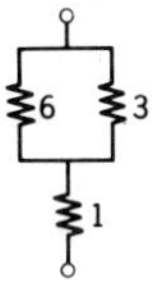

The total resistance of this circuit is _____ ohms. 12

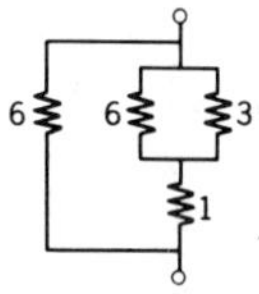

The total resistance of this circuit is _____ ohms. 84

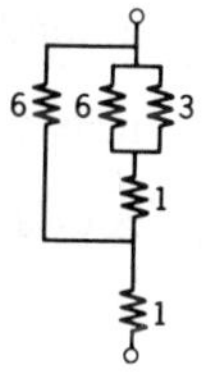

The total resistance of this circuit is _____ ohms. 97

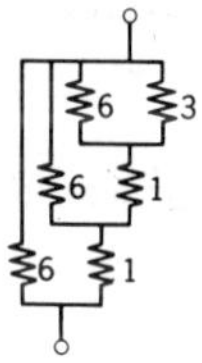

The total resistance of this circuit is _____ ohms. 78

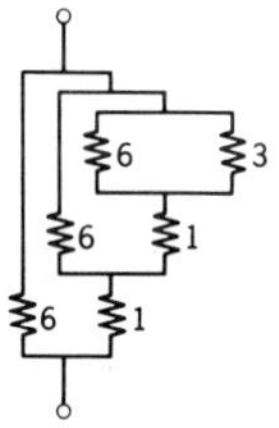

The total resistance of this circuit is _____ ohms. 10

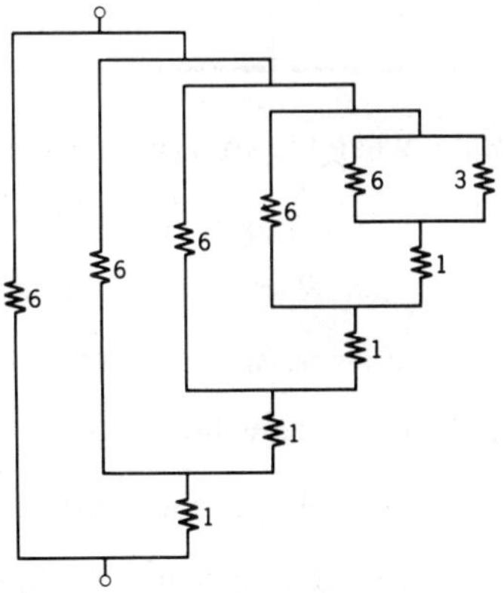

Go to 9-14.

9-13
(from 9-3)

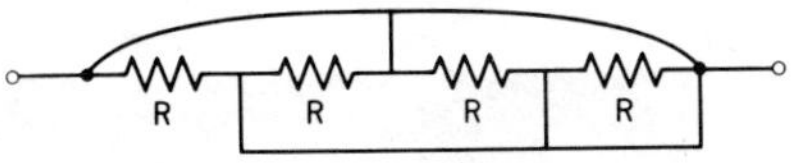

The total resistance is zero; i.e., a short circuit [*true* ; *false*]. 33

Go to 9-12

9-14 The power in watts dissipated in a resistor R is IV where I is in [coulombs ; amperes ;   91
(from statcoulombs] and V is in [volts ; statvolts] .   43
9-12)
The power also is $\left[IR^2 \; ; \; \dfrac{R^2}{I} \; ; \; IR \; ; \; I^2R \; ; \; \text{none of these} \right]$. It also has the value   7

$\left[V^2R \; ; \; VR^2 \; ; \; \dfrac{V^2}{R} \; ; \; \dfrac{V}{R^2} \; ; \; \text{none of these} \right]$.   24

If incorrect go to 9-20 ; if correct go to 9-15.

9-15 The next three frames deal with diodes. We shall assume ideal diodes; i.e., when the diode
(from is conducting it offers zero resistance, and when it has a reverse potential, it offers infinite
9-14 or
9-20) resistance.

A plot of the current I vs. voltage V across a diode would look

like Curve _____ . A plot of the current vs. voltage for an   16

ordinary resistor would look like Curve _____ .   57

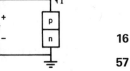

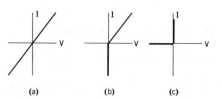

(a)   (b)   (c)

Go to 9-16

9-16 When the voltage at A is +5 volts the diode will conduct
(from [*true* ; *false*] . If the current through the diode is I,   71
9-15)
then $IR =$ _____ volts where R is the resistance of the   23

series resistor. The voltage drop across the diode is _____ volts.   50

When the voltage at A is −5 volts the diode will conduct [*true* ; *false*] .   101

The voltage drop across the resistor would be $IR =$ [0 ; 5] volts. The   61

voltage drop across the diode would be [0 ; 5] volts.   82

Go to 9-17

132

9-17
(from 9-16)

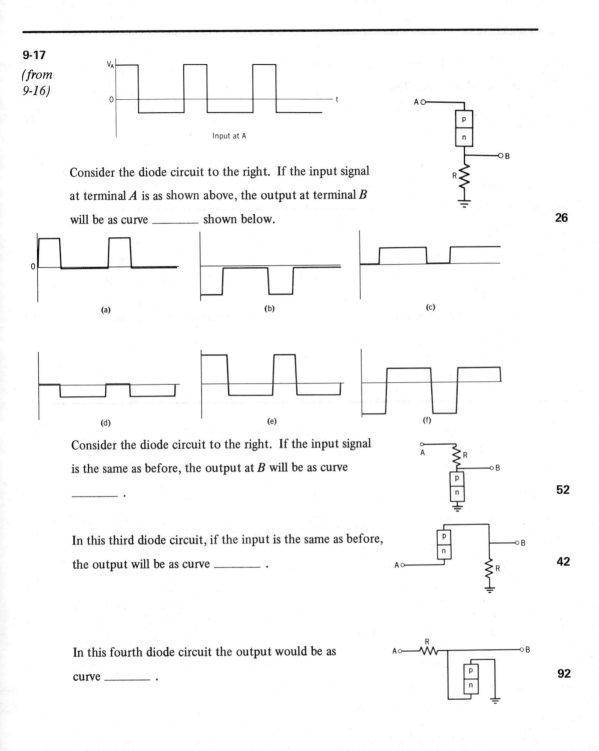

Consider the diode circuit to the right. If the input signal at terminal A is as shown above, the output at terminal B will be as curve _____ shown below. 26

Consider the diode circuit to the right. If the input signal is the same as before, the output at B will be as curve _____ . 52

In this third diode circuit, if the input is the same as before, the output will be as curve _____ . 42

In this fourth diode circuit the output would be as curve _____ . 92

If incorrect go to 9-21 ; if correct go to 9-18.

9-18
(from 9-17 or 9-21)
An RF voltage is heavily modulated with an audio frequency sine wave. The curve represents $\left[\dfrac{1}{2} \; ; \; 1 \; ; \; 1\dfrac{1}{2} \; ; \; 2 \; ; \; 4\right]$ periods of the audio frequency.

20

If incorrect go to 9-22 ; if correct go to 9-23.

9-19
(from 9-23)

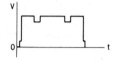

The above video signal for a single scan line would give a scan line looking like [a ; b ; c] .

39

Go to 9-26.

9-20
(from 9-14)
Power = $I(V)$ where $V = \left[IR \; ; \; \dfrac{I}{R} \; ; \; \dfrac{R}{I}\right]$ and power = $V(I)$ where $I = \left[\dfrac{V}{R} \; ; \; VR \; ; \; \dfrac{R}{V}\right]$.

104
31

Go to 9-15.

9-21
(from 9-17)

Let us analyze the second circuit in more detail. Suppose a positive voltage V_1 is applied at A. The diode will experience a [forward ; reverse] potential. The diode will be [conducting ; non-conducting] and have essentially [zero ; infinite] resistance. The voltage drop across the diode will be [0 ; V_1] and the voltage drop across R will be [0 ; V_1] .

Now apply a negative voltage $(-V_2)$ to A. Now the diode will have essentially [zero ; infinite] resistance. The voltage drop across R will be [0 ; V_2] and the voltage drop across the diode will be [0 ; $-V_2$] .

Suppose a positive voltage V_1 followed by a negative voltage $-V_2$ is applied to A. The output voltage at B would look like curve [a ; b ; c ; none of these] .

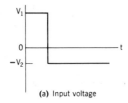
(a) Input voltage

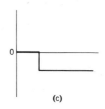
(b)   (c)

Go to 9-18.

9-22
(from 9-18)

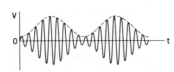

If the above signal is rectified by a diode detector, the output voltage will be as shown. The time interval $(t_2 - t_1)$ is $\left[\dfrac{1}{2} \; ; \; 1 \; ; \; 2\right]$ complete sinewaves.

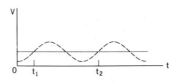

Go to 9-23.

9-23
(from 9-18 or 9-22)

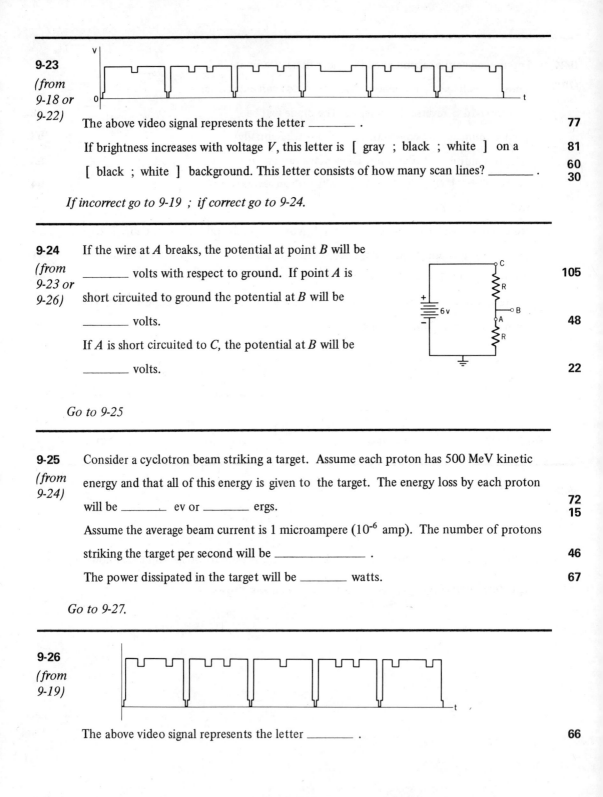

The above video signal represents the letter _____ .   77

If brightness increases with voltage *V*, this letter is [gray ; black ; white] on a   81

[black ; white] background. This letter consists of how many scan lines? _____ .   60 / 30

If incorrect go to 9-19 ; if correct go to 9-24.

9-24
(from 9-23 or 9-26)

If the wire at *A* breaks, the potential at point *B* will be _____ volts with respect to ground. If point *A* is short circuited to ground the potential at *B* will be _____ volts.   105 / 48

If *A* is short circuited to *C*, the potential at *B* will be _____ volts.   22

Go to 9-25

9-25
(from 9-24)

Consider a cyclotron beam striking a target. Assume each proton has 500 MeV kinetic energy and that all of this energy is given to the target. The energy loss by each proton will be _____ ev or _____ ergs.   72 / 15

Assume the average beam current is 1 microampere (10^{-6} amp). The number of protons striking the target per second will be _____ .   46

The power dissipated in the target will be _____ watts.   67

Go to 9-27.

9-26
(from 9-19)

The above video signal represents the letter _____ .   66

Go to 9-24.

9-27
(from 9-25)

In the next two frames we will show that the time for a proton to make one complete circle in a cyclotron is independent of the velocity of the proton or the radius of its path.

We start with $F = Ma$ where $a = \dfrac{v^2}{R}$ is the acceleration. The net force F is the

[magnetic force ; emf ; electrostatic force] acting on the moving charged particle.   89

$F = \left[\dfrac{evB}{c} \; ; \; \dfrac{e^2}{R^2} \; ; \; \text{neither} \right]$.   74

Now equate the above two expressions for F and divide both sides by v. The result is

$M\dfrac{v}{R} = \dfrac{eB}{c}$ [*true* ; *false*] .   55

The quantity $\dfrac{v}{R}$ is $\left[\dfrac{2\pi}{T} \; ; \; \dfrac{1}{T} \; ; \; 2\pi T \; ; \; \dfrac{T}{2\pi} \; ; \; \text{none of these} \right]$, which is independent of v.   9

Go to 9-28.

9-28
(from 9-27)

Substituting $\dfrac{2\pi}{T}$ for $\dfrac{v}{R}$ in the previous frame gives $M\dfrac{2\pi}{T} = \dfrac{eB}{c}$ and $T = \dfrac{2\pi Mc}{eB}$

[*true* ; *false*] . As long as M is independent of v, we have the result that the period T 44

is independent of v. According to the theory of relativity $M = \dfrac{M_0}{\sqrt{1 - (v^2/c^2)}}$

Then if v gets close to c, the period T [increases ; decreases] and the cyclotron   28

oscillator frequency must be [increased ; decreased] as the particle energy increases.   51

Go to 9-29.

9-29
(from 9-28)

In this problem we shall be given the dimensions of an electron synchroton. From these dimensions we shall be able to calculate the energy of this high energy accelerator.

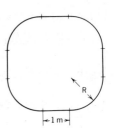

This synchrotron contains 4 straight sections of 1 meter length each. The total length of path around the synchroton is 12 meters. The peak value of the magnetic field is 10,000 gauss.

The total length of the curved sections is then _____ meters.   83

The radius of curvature of the curved sections is _____ cm. The peak magnetic field   100

in the straight sections must be _____ gauss. The low energy electrons must be in-   108

jected when the magnetic field is at its [minimum ; maximum] . When in the curved   49

sections the force on the electrons is $\frac{mv^2}{R}$ [true ; false] . When at their maximum   54

energy the velocity v is for all practical purposes equal to c. To this approximation, the

force on the electrons is $\left[E \ ; \ eB \ ; \ B \ ; \ \frac{eB}{c} \right]$ and the centripetal force is $\frac{mc^2}{R}$. This gives   69

the result that $mc^2 = eBR$ [true ; false] . We will learn in Chapter 11 that mc^2 is the   59

total energy of a particle. Hence $W = eBR$.

Go to 9-30.

9-30
(from 9-29)

Putting in the numbers gives

$W = 4.8 \times 10^{-10} \times 10{,}000 \times \dfrac{400}{\pi}$ [watts ; joules ; ergs ; ev ; none of these] .   1

$W = $ _____ ergs.   87

$W = $ _____ Mev.   18

If the rest energy of an electron is 0.5 Mev, the kinetic energy in this problem is

$KE = $ [37.7 ; 381.5 ; neither] Mev.   98

Go to 9-31.

9-31 Review Questions:

(from 9-30)

1. We know that the resistance of a metal increases with increasing temperature. Does the mean free path of the conduction electrons increase or decrease with increasing temperature? _____ **2**

2. How many watts are dissipated in a 60 ohm light bulb connected across 120 volts?

_____ watts. **79**

3. Which will dissipate more power when connected across a 6 volt battery, two resistors in series, or the same two in parallel? _____ **37**

4. A 10 Mev electron has a radius of curvature of 10 cm when in a certain magnetic field. In the same field what will be the radius of curvature of an electron with twice the momentum? _____ cm. Will this second electron have twice the velocity? _____ **63 65**

5. What is the voltage at A? $V_A =$ _____ volts. **103**

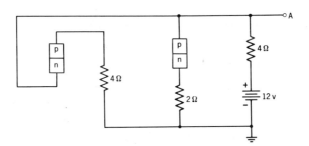

Chapter 9 Answer Sheet

Item Number	Answer
1	ergs
2	decrease
3	parallel
4	3×10^2
5	same
6	infinite
7	I^2R
8	3
9	$\dfrac{2\pi}{T}$
10	2
11	false
12	2
13	$\eta e \dfrac{\Delta u}{2}$
14	$\dfrac{V}{L}$
15	8×10^{-4}
16	c
17	$\dfrac{1}{\dfrac{1}{R_1}+\dfrac{1}{R_2}+\dfrac{1}{R_3}}$
18	382
19	c
20	2
21	inversely
22	6
23	5
24	$\dfrac{V^2}{R}$
25	V_1
26	a
27	true
28	increases
29	directly
30	6
31	$\dfrac{V}{R}$
32	2
33	true
34	$\eta A v_d$
35	2
36	1
37	parallel
38	halved
39	b
40	R_1
41	zero
42	d
43	volts
44	true
45	$\dfrac{\eta e^2 L}{2mu}$
46	$\dfrac{10^{-6} \times 3 \times 10^9}{4.8 \times 10^{-10}}$
47	iA
48	zero
49	minimum
50	zero
51	decreased
52	d
53	conducting
54	true
55	true
56	$\dfrac{R_3 R_2 + R_1 R_3 + R_2 R_1}{R_1 R_2 R_3}$
57	a
58	zero
59	true
60	white
61	zero
62	neither
63	20 cm
64	$\dfrac{EL}{iA}$
65	no
66	X
67	500
68	forward
69	eB
70	$\dfrac{i}{E}$
71	true
72	5×10^8
73	1 erg
74	$\dfrac{evB}{c}$
75	$\dfrac{V}{I}$
76	$I_1 - I_3$
77	A
78	2
79	240 watts
80	$\dfrac{L}{\sigma A}$
81	gray
82	5
83	8
84	3
85	$\eta e \bar{v}_d$
86	$\eta A e$
87	6.1×10^{-4}
88	true
89	magnetic force
90	$\dfrac{\Delta u}{2}$
91	amperes
92	a
93	$-V_2$
94	zero
95	false
96	1 joule
97	2
98	381.5
99	$\dfrac{R_1 R_2}{R_1 + R_2}$
100	$\dfrac{400}{\pi}$
101	false
102	$\eta A \bar{v}_d e$
103	4 volts
104	IR
105	6
106	true
107	$\dfrac{10}{3 \times 10^{10}}$
108	zero

Chapter 10
WAVE MOTION AND LIGHT

10-1 An electromagnetic wave is traveling to the right in the x-direction. At time $t = 0$, the plot of E vs x looks like curve (a). At time $t = \frac{1}{4}T$, the plot of E vs x would be like curve _____ . At a time $t = \frac{1}{2}T$ the plot would be like curve _____ . At a time $t = -\frac{1}{4}T$ the plot would look like curve _____ .

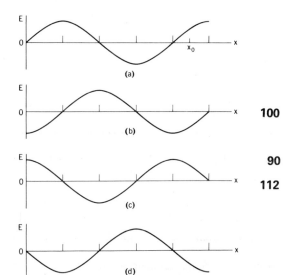

100

90
112

Go to 10-2.

10-2 At a given position $x = x_0$, the value of E is zero [once; twice ; four times] during each period T. The number of wave fronts or crests of maximum wave amplitude to pass x_0 in one second is $\left[T ; \frac{1}{T} ; 2T ; \frac{2}{T} ; \text{none of these} \right]$.

(from 10-1)

81

48

Wave intensity is proportional to the [square root ; square ; cube] of the amplitude. The number of wave fronts of maximum intensity to pass x_0 in one second is $\left[T ; 2T ; \frac{1}{T} ; \frac{2}{T} ; \text{none of these} \right]$.

66

13

Go to 10-10.

143

10-3
(from 10-10)

If curve a is the plot of the amplitude of a wave vs x, curve _____ will be a plot of the intensity vs x. The distance from A to _____ is one wavelength λ.

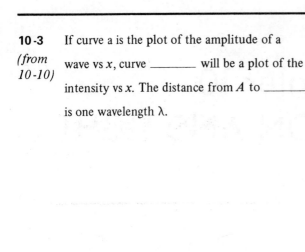

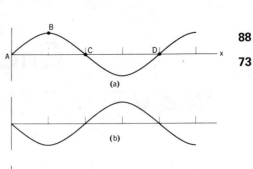

Go to 10-4.

10-4
(from 10-3)

The period of oscillation of an electromagnetic wave is $T = \left[\dfrac{\lambda}{c}\ ;\ \dfrac{c}{\lambda}\ ;\ \lambda c\ ;\ \dfrac{1}{\lambda c}\ ;\ \text{none of these}\right]$.

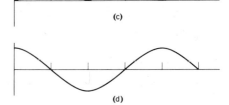

If incorrect go to 10-11 ; if correct go to 10-5.

10-5
(from 10-4 or 10-11)

The frequency range of visible light is from 4.3×10^{14} cps (cycles per second) to 7×10^{14} cps. Electromagnetic radiation of 7×10^{14} cps appears as [red ; blue ; violet ; none of these] light. The number of wavelengths of red light in one cm is _____ . The number of wavelengths in one cm is $\left[f\ ;\ \dfrac{1}{f}\ ;\ \dfrac{c}{f}\ ;\ \dfrac{f}{c}\ ;\ fc\ ;\ \dfrac{1}{fc}\ ;\ \text{none of these}\right]$.

3
39
20

Go to 10-6.

10-6
(from 10-5)

One end of a string is vibrated at a constant f. As the tension of the string increases, the wavelength λ on the string will [increase ; decrease] .

52

If incorrect go to 10-14 ; if correct go to 10-7.

10-7
(from 10-6 or 10-14)

Consider a single continuous sine wave moving along a string. Each point on the string is moving up and down in SHM (simple harmonic motion) [*true* ; *false*].   67

Each point on the string moves up and down with the same maximum displacement [*true* ; *false*]. In each period of oscillation the entire string will have zero displacement   110
[once ; twice ; four times ; never] .   33

Go to 10-8.

10-8
(from 10-7)

Consider a standing wave on a string with four loops ($N = 4$). Each point on the string is moving up and down in SHM [*true* ; *false*]. Each point on the string moves up and   12
down with the same maximum displacement [*true* ; *false*]. The length of the string   97
is _____ wavelengths.   54

During each period of oscillation the entire string will have zero displacement   101
[once ; twice ; four times ; never] .

Go to 10-9.

10-9
(from 10-8)

Consider three successive positions, A, B, and C, of the string in the previous frame. The time difference between positions A and C is $\left[\dfrac{1}{4} ; \dfrac{1}{2} ; 1 ; 2\right]$ times T, the period of oscillation. Not counting the ends, the total number of nodes is _____ .

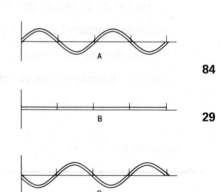

84

29

Go to 10-12.

10-10
(from 10-2)

If T is the period, the number of complete oscillations in t seconds will be _____ .   28
The number of complete oscillations in one second will be $f = \dfrac{1}{T}$ [*true* ; *false*].   69

Go to 10-3.

10-11
(from 10-4)

The frequency f is $f = \left[\dfrac{c}{\lambda} \; ; \; \dfrac{\lambda}{c} \; ; \; \lambda c \; ; \; \dfrac{1}{\lambda c} \; ; \text{ none of these} \right]$. 42

The period $T = \left[\dfrac{1}{f} \; ; \; \dfrac{c}{f} \; ; \; fc \; ; \text{ none of these} \right]$. 8

Substituting the value for f in the previous result gives $T =$ _____ . 16

Go to *10-5*.

10-12
(from 10-9)

Two sine waves are traveling in opposite directions along a string. If the two sine waves are in phase at $t = 0$, the corresponding crests of the two will be $\left[\dfrac{1}{8} \; ; \; \dfrac{1}{4} \; ; \; \dfrac{1}{2} \; ; \; 1 \right]$ wavelength apart at $t = \dfrac{1}{4}T$ where T is the period. 17

At time $t = \dfrac{1}{2}T$ the displacements on the string will be [maximum ; minimum ; neither] . 92

If incorrect go to *10-30* ; if correct go to *10-15*.

10-13
(from 10-16)

A microwave oscillator emits plane electromagnetic waves to the right which are reflected back to the left. P_1 and P_2 are the positions of two successive intensity minima. If the distance from P_1 to P_2 is 4 cm, the wavelength is $\lambda =$ _____ cm. The frequency of the oscillator is $f =$ _____ cps. 87

56

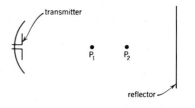

Go to *10-17*.

10-14
(from 10-6)

As the tension of the string increases, the wave velocity [increases ; decreases] . 36

The relation between wavelength and wave velocity is $\lambda = \left[\dfrac{v}{f} \; ; \; fv \; ; \; \dfrac{f}{v} \; ; \text{ none of these} \right]$. 4

As the velocity increases the wavelength [increases ; decreases] . 43

Go to *10-7*.

146

10-15
(from 10-12 or 10-30)

If S_1 and S_2 are vibrating in phase, the condition for a maximum is $D_1 - D_2 =$ _____ . The condition for a minimum is $D_1 - D_2 =$ _____ .
In the above answers N could be zero
[*true* ; *false*] .

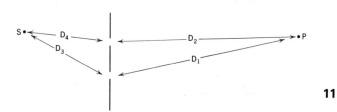

41

89

38

Go to 10-16.

10-16
(from 10-15)

If S_1 and S_2 are vibrating out of phase (when S_1 moves up, S_2 moves down), a point equidistant from S_1 and S_2 will be on a line of nodes [*true* ; *false*] .
Now the condition for a minimum will be $D_1 - D_2 =$ _____ and the condition for a maximum will be $D_1 - D_2 =$ _____ .

108

94

45

Go to 10-13.

10-17
(from 10-13)

The condition for a double slit interference at P (P is on a line of nodes) is $D_1 - D_2 = \left(N + \frac{1}{2}\right)\lambda$ [*true* ; *false*] .

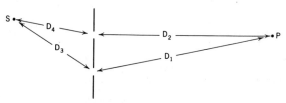

11

Go to 10-18.

10-18
(from 10-17)

The condition for a destructive interference at P is $\left[D_1 - D_2 = N\lambda \; ; \; D_3 - D_4 = \left(N + \frac{1}{2}\right)\lambda \; ; \; D_1 + D_3 - D_2 - D_4 = \left(N + \frac{1}{2}\right)\lambda \; ; \; (D_1 - D_2) + (D_3 - D_4) = N\lambda \right]$.

104

If incorrect go to 10-23 ; if correct go to 10-19.

10-19
(from 10-18 or 10-23)

The condition for an intensity maximum on the screen is $\sin\theta = N\lambda/d$ [*true* ; *false*] .

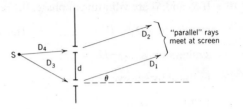

75

Go to 10-21.

10-20
(from 10-29)

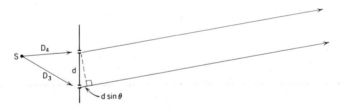

The rays from pt. A will strike the convex mirror and be reflected. The reflected rays will all appear to come from a common point in space. This common point is defined as the image of A. The image of A will be at [P ; F ; C ; none of these] . It will be [to the left of P ; between P and F ; between F and C ; to the right of C] . The image point will be [above ; below ; on] the optic axis (line BC).

57
85
107

Go to 10-31.

10-21
(from 10-19)

The path difference from S to the screen is [$D_3 - D_4$; $d\sin\theta$; $D_3 - D_4 + d\sin\theta$; $D_3 - D_4 - d\sin\theta$; none of these] .

1

In order to have a minimum at the screen this path difference must be _____ .

5

The condition on θ for a maximum is $\sin\theta = \left[\dfrac{N\lambda}{d} \; ; \; \dfrac{(N\lambda + D_4 - D_3)}{d} \; ; \; (N\lambda - D_4 + D_3)d \; ; \; \dfrac{N\lambda}{d} + D_4 - D_3 \right]$.

50

Go to 10-24.

10-22
(from 10-27)

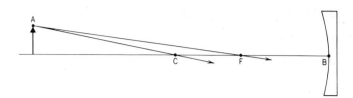

Consider a concave mirror having pt. C as the center of curvature and pt. F as the focus.

When the ray from A to C strikes the mirror, the reflected ray will [pass through F ; pass through C ; be parallel to CB] . 24

When the ray from A to F strikes the mirror, the reflected ray will [pass through F ; pass through C ; be parallel to CB] . 60

The length FB is _____ of CB. 86

Go to 10-29.

10-23
(from 10-18)

The path difference is [$D_1 - D_2$; $D_3 - D_4$; $D_1 + D_3 - D_2 - D_4$; none of these] . 9

There will be an intensity [minimum ; maximum] if the path difference is is $\left(N + \dfrac{1}{2}\right)\lambda$. 46

Go to 10-19.

10-24
(from 10-21)

If y is the displacement of a spring, then the spring force on M is $F = -ky$ where k is the force constant of the spring [*true* ; *false*].
The work done in stretching the spring from $y = 0$ to $y = y_0$ is the potential energy of the spring when $y = y_0$ [*true* ; *false*]. The potential energy $U = \bar{F} \cdot y_0$ where $\bar{F}$ is the average force exerted over the distance y_0 [*true* ; *false*]. The magnitude of the average force is $\bar{F} =$ $\left[ky_0 \;\; ; \;\; \dfrac{1}{2}ky_0 \;\; ; \;\; \text{neither} \right]$. Then $U =$ _____ times ky_0^2.

If the spring is stretched at a distance y_0 and then released, M will oscillate in SHM and the total mechanical energy of M will be _____. We make the general observation that in SHM the energy (or intensity) is proportional to the [first power ; square ; inverse square] of the displacement. Wave motion is also SHM [*true* ; *false*]. In wave motion the intensity is proportional to the square of the displacement (or amplitude) [*true* ; *false*].

18
15
49
63
59
25
21
10
6

Go to 10-25.

10-25
(from 10-24)

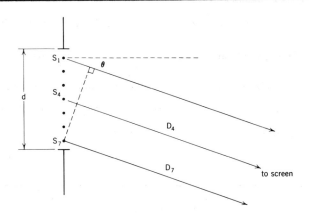

Consider seven evenly spaced monochromatic sources of light $S_1, S_2, \ldots, S_7$ all in phase. At the screen the light from S_4 will cancel S_7 if $D_4 - D_7 = $ _____, or if $\frac{d}{2}\sin\theta =$ _____.

The light from S_3 will cancel S_6 if $\frac{d}{2}\sin\theta =$ _____. The light from S_2 will cancel _____ if $d\sin\theta = \lambda$.

At this same angle, all the light reaching the screen is cancelled out [*true* ; *false*].

The condition for the above intensity minimum is $\sin\theta =$ _____.

This phenomenon is called single slit diffraction and occurs whenever light waves strike a single slit as shown below.

14
82
106
98
55
77

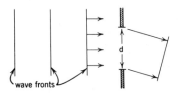

Go to 10-26.

10-26
(from 10-25)

If two of the previous slits are put side by side, and the angle θ is such that $\sin\theta = \frac{\lambda}{d}$, there will be an intensity minimum from each of the two slits [*true* ; *false*]. There would also be an intensity minimum from a double-sized slit having a width of $2d$ [*true* ; *false*].

71
2

Go to 10-27.

10-27
(from 10-26)
Consider a single slit of width d. Let us divide d into N equal intervals. Then each interval can be considered as a single slit of width $d' = \left[Nd \; ; \; \dfrac{d}{N} \; ; \; d \right]$. When $\sin\theta = \dfrac{\lambda}{d'}$ each interval will give cancellation. Hence there is an intensity minimum coming from the entire slit width d. Then $\sin\theta = $ _____ is the general condition for the Nth minimum of a single slit diffraction pattern.

35

80

Go to 10-22.

10-28
(from 10-34)
A source of light is imbedded in a piece of glass of index of refraction $n = 1.5$. The angle of refraction in the air is θ_2. θ_2 will always be [greater ; less] than θ_1. The relation is
$\left[\dfrac{\theta_1}{\theta_2} = \dfrac{1}{n} \; ; \; \dfrac{\theta_1}{\theta_2} = n \; ; \; \dfrac{\sin\theta_1}{\sin\theta_2} = \dfrac{1}{n} \; ; \; \dfrac{\sin\theta_1}{\sin\theta_2} = n \; ; \; \dfrac{\cos\theta_1}{\cos\theta_2} = n \right]$.

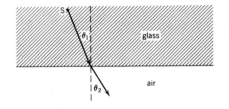

47

93

Go to 10-32.

10-29
(from 10-22)

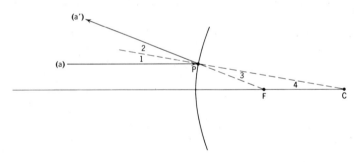

Consider a convex spherical mirror with pt. C as the center of curvature. Parallel ray (a) is reflected as ray (a').

Parallel rays striking the convex mirror appear to come from [F ; C ; P]. Point [F ; C ; neither] is the focus. Angle 1 is exactly equal to angle 2 [true ; false]. This is because of the law of [reflection ; refraction]. Angle 1 is equal to angle 4 [true ; false]. Angle 1 is equal to angle 3 [true ; false]. Triangle PFC is isoscles [true ; false]. Side PF equals FC [true ; false]. The distance FC is almost one half the radius of the circle (for small angles) [true ; false].

111
95
51
23
19
83
62
65
7

Go to 10-20.

152

10-30
(from 10-12)

In $\frac{1}{4}$ of a period one wave will move $\left[\frac{1}{8}\ ;\ \frac{1}{4}\ ;\ \frac{1}{2}\right]$ wavelength to the right and the other wave will move $\left[\frac{1}{8}\ ;\ \frac{1}{4}\ ;\ \frac{1}{2}\right]$ wavelength to the left. The relative separation of the two waves will be $\left[\frac{1}{8}\ ;\ \frac{1}{4}\ ;\ \frac{1}{2}\ ;\ 1\right]$ wavelength.

37

22

96

Go to 10-15.

10-31
(from 10-20)

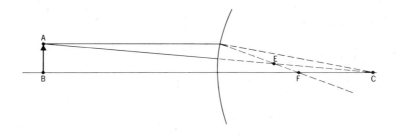

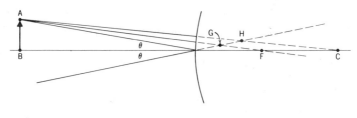

The image of A will be at pt. E [*true* ; *false*]. The image of A will be at [G ; H ; neither].

99

102

Go to 10-33.

10-32
(from 10-28)

Suppose θ_1 is so large that θ_2 is almost $90°$.

Then θ_1 would have the value given by

$\sin\theta_1 = \left[n\ ;\ \dfrac{1}{n}\ ;\ \text{neither}\right]$.

The value of $\sin\theta_1$ would be [1.5 ; 0.667 ; neither].

26

61

This means that light coming from S having angles of incidence greater than $42°$ will not leave the glass, but will instead be totally reflected back into the glass. This phenomenon is called total internal reflection.

Go to 10-36.

10-33
(from 10-3)

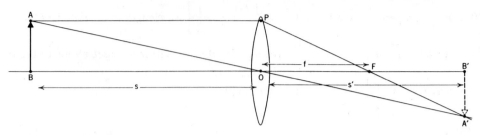

We would like to derive an algebraic relation between object distance s and image distance s', and the focal length f.

Triangle ABO is similar to triangle [OPF ; $A'B'F$; $A'B'O$].   44

From these two similar triangles, the ratio $\dfrac{A'B'}{AB} = \left[\dfrac{s'}{f} \ ; \ \dfrac{s'}{s} \ ; \ \dfrac{s'-f}{s} \ ; \ \dfrac{f}{s} \ ; \ \dfrac{s}{f} \right]$.   70

Triangle POF is similar to triangle _____ .   64

From these two triangles the ratio $\dfrac{A'B'}{AB} = \left[\dfrac{s'-f}{f} \ ; \ \dfrac{f}{s'-f} \ ; \ \dfrac{s'}{f} \ ; \ \dfrac{f}{s'} \right]$ 78

The length $AB =$ _____ .   27

The right hand sides of Eqs. (1) and (2) must be equal [*true* ; *false*].   79

Equate these right hand sides and solve for $\dfrac{1}{f}$:

$\dfrac{1}{f} = \left[\dfrac{1}{s} - \dfrac{1}{s'} \ ; \ \dfrac{1}{s'} - \dfrac{1}{s} \ ; \ \dfrac{1}{s} + \dfrac{1}{s'} \ ; \ \dfrac{ss'}{s+s'} \ ; \ \text{none of these} \right]$.   72

Go to 10-35.

10-34
(from 10-35)

If s is greater than $2f$, s' will be [less ; greater] than $2f$ and the image will be   30
[larger ; smaller] than the object.   40

If s is between f and $2f$, s' will be [less ; greater] than $2f$ and the image will be   109
[larger ; smaller] than the object.   34

Go to 10-28.

10-35
(from 10-33)

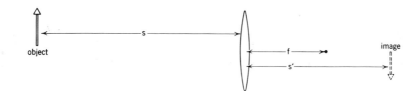

The relation between image and object distance is $\frac{1}{s} + \frac{1}{s'} = \frac{1}{f}$ and is called the thin lens formula.

There is a sign convention to tell whether s, s', or f is positive or negative. According to this convention the problem must always be set up so that the light passes through the lens from left to right. Then s' will be positive if the [object ; image] is to the [left ; right] of the lens. The quantity f is negative if the lens is a [converging ; diverging] lens.

74
103
32

If the rays passing through the lens are converging on a virtual object (it could be the image of a preceding lens to the left), then s will be [positive ; negative] . **68**

Go to 10-34.

10-36 Review Questions:

(from 10-32)

1. What is the wavelength in terms of the period T and the wave velocity v?
$$\lambda = \underline{}$$ **114**

2. If n crests of a wave are received per minute, what is the period of the wave in seconds?
$$T = \underline{}$$ **105**

3. How many wavelengths long is a vibrating string having 3 nodes (not counting the two ends as nodes)? $L = \underline{}$ **31**

4. An object is at a distance of 8 inches from a concave mirror having a 12 inch radius of curvature
   a. What is the focal length? $F = \underline{}$ **76**
   b. Would the image be inverted?   _____ **113**
   c. Where would the image be located?   _____ **115**

Chapter 10 Answer Sheet

Item Number	Answer	Item Number	Answer	Item Number	Answer
1	$D_3 - D_4 + d\sin\theta$	39	1.43×10^4	81	twice
2	true	40	smaller	82	$\frac{1}{2}\lambda$
3	violet	41	$N\lambda$	83	true
4	$\frac{v}{f}$	42	$\frac{c}{\lambda}$	84	$\frac{1}{2}$
5	$\left(N + \frac{1}{2}\right)\lambda$	43	increases	85	between P and F
6	true	44	$A'B'O$	86	$\frac{1}{2}$
7	true	45	$\left(N + \frac{1}{2}\right)\lambda$	87	8
8	$\frac{1}{f}$	46	minimum	88	c
9	$D_1 + D_3 - D_2 - D_4$	47	greater	89	$\left(N + \frac{1}{2}\right)\lambda$
10	true	48	$\frac{1}{T}$	90	d
11	false	49	true	91	$\frac{\lambda}{c}$
12	true	50	$\frac{N\lambda + D_4 - D_3}{d}$	92	maximum
13	$\frac{2}{T}$	51	true	93	$\frac{\sin\theta_1}{\sin\theta_2} = \frac{1}{n}$
14	$\frac{1}{2}\lambda$	52	increase	94	$N\lambda$
15	true	53	neither	95	F
16	$\frac{\lambda}{c}$	54	2	96	$\frac{1}{2}$
17	$\frac{1}{2}$	55	true	97	false
18	true	56	3.75×10^9	98	S_5
19	true	57	none	99	true
20	$\frac{f}{c}$	59	$\frac{1}{2}$	100	b
21	square	60	be parallel to CB	101	twice
22	$\frac{1}{4}$	61	0.667	102	H
23	reflection	62	true	103	right
24	pass through C	63	$\frac{1}{2}ky_0$	104	$D_1 + D_3 - D_2 - D_4 = \left(N + \frac{1}{2}\right)\lambda$
25	$\frac{1}{2}ky_0^2$	64	$A'B'F$		
26	$\frac{1}{n}$	65	true	105	$\frac{120}{n}$
27	PO	66	square	106	$\frac{1}{2}\lambda$
28	$\frac{t}{T}$	67	true	107	above
29	3	68	negative	108	true
30	less	69	true	109	greater
31	2	70	$\frac{s'}{s}$	110	true
32	diverging	71	true	111	F
33	never	72	$\frac{1}{s} + \frac{1}{s'}$	112	c
34	larger	73	D	113	yes
35	$\frac{d}{N}$	74	image	114	vT
36	increases	75	false	115	24 inches in front of the mirror
37	$\frac{1}{4}$	76	6 inches		
38	true	77	$\frac{\lambda}{d}$		
		78	$\frac{s' - f}{f}$		
		79	true		
		80	$\frac{N\lambda}{d}$		

Chapter 11
RELATIVITY

11-1 According to the ether theory of light, if the light source is at rest with respect to the ether, the speed of light will be $v = c$ and will be independent of the velocity of the observer [*true* ; *false*].   **88**

According to the ether theory if the observer is at rest with respect to the ether, the speed of light will be $v = c$ even if the source is moving with respect to the ether [*true* ; *false*].   **45**

According to the ether theory, Maxwell's equations would be true only in one reference system; namely, the system at rest with respect to the ether [*true* ; *false*].   **51**

Go to 11-2

11-2
(from 11-1)
A boat makes a round trip between two towns 60 miles apart on a river. If the boat's speed is 20 mph and the water is not flowing, the time required to make the round trip is _____ hours. If the river is flowing at 10 mph, the time to travel the 60 miles downstream is _____ hours and the time to travel the 60 miles upstream is _____ hours. When the river is flowing the round trip takes _____ hours [more ; less] than when the river is not flowing.   **72 79 18 24 38**

When the ether is flowing along the arm of an interferometer, the time for light to make the round trip would be [more ; less] than the time when the ether is at rest, for exactly the same reasons.   **59**

Go to 11-3.

11-3
(from 11-2)

Suppose two small islands are 60 mi apart and a boat is to make a round trip between them. If the current is 10 mph at right angles to the line joining the two islands, and the boat's velocity with respect to the water is 20 mph as in the previous frame, then the boat's velocity with respect to the islands will be

_____ mph. The time to make the round trip is approximately [3 ; 6 ; 7 ; 8 ; 9] hours (choose the nearest to the correct answer).

This result is [shorter than ; longer than ; the same as] the time for the round trip when the same current is parallel to the path. The same reasoning holds in the ether theory. When the arm is rotated 90° so that it is perpendicular rather than parallel to the ether flow, then the time for light to make the round trip will be [greater ; less] .

1
28

16

78

Go to 11-4

11-4
(from 11-3)

A pulse of light travels the round trip from mirror M_1 to M_2 and back to M_1. If the arm containing the mirrors is moving through the ether with velocity v, the time for the round trip will be [greater ; less] than when the arm is at rest with respect to the ether.

If the ether is moving parallel to the arm with velocity v, the time for the round trip will be [greater ; less] than when the ether is moving perpendicular to the arm with the same velocity v.

33

61

Go to 11-5.

11-5
(from 11-4)

From now on we shall discard the ether theory and instead use the now accepted Einstein theory of relativity.

Consider observer A as stationary. At $t = 0$ a pulse of light is emitted at point O. The time for the light to go a distance R is

$t = \dfrac{R}{c}$ [true ; false]. If the source of light has been moving to the right when the pulse was emitted at O, the time to travel the distance R would have been [less than ; more than ; the same as] the previous answer. The position x of the pulse at a time t is $x = ct$ [true ; false]. Now consider observer B who is moving at velocity v to the right. We will use primed coordinates x' and t' for the distance and time measurements made by B. At the time $t = 0$ when the light pulse was emitted, both observers were passing each other at point O. According to A the position of B at time t is $x = \left[vt \ ; \ -vt \ ; \ \dfrac{1}{\sqrt{1-(v/c)^2}} \ ; \ 0 \ ; \ \text{none of these} \right]$.

According to B the position of A is $x' = \left[vt' \ ; \ -vt' \ ; \ \dfrac{-vt'}{\sqrt{1-(v/c)^2}} \ ; \ 0 \ ; \ \text{none of these} \right]$.

According to B the position of the light pulse will be $x' = [\ (c+v)t \ ; \ (c-v)t \ ; \ ct' \ ; \ \text{none of these} \ ]$.

How can both equations $x = ct$ and $x' = ct'$ be satisfied by the very same light pulse? The answer is that the relation between x and x' is not $x' = x - vt$, but it is $x' = \gamma(x - vt)$ where $\gamma = 1/\sqrt{1-(v/c)^2}$. Likewise the time relation is not the classical $t = t'$, but it is $t' = \gamma(t - vx/c^2)$. The reverse transformation is $x = \gamma(x' + vt')$ and $t = \gamma\left(t' + \dfrac{vx'}{c^2}\right)$. In fact these Lorentz transformations as they are called, were just so designed to make the equation $x' = ct'$ hold true whenever the equation $x = ct$ also holds true. We will now show this.

Go to 11-6.

11-6
(from 11-5)

This is a continuation of the discussion in frame 11-5.

Given: According to the stationary observer the position of the light pulse is $x = ct$.

To prove: For the moving observer B, the position of the same light pulse will be
$x' = ct'$ assuming the relations between x' and x, and t' and t are given by the Lorentz transformations.

In this proof we start with

$$x = ct$$

The left hand side is $[\ x' - vt'\ ;\ x' + vt'\ ;\ \gamma(x' - vt')\ ;\ \gamma(x' + vt')\ ]$. 10

The right hand side is $\left[\ ct'\ ;\ \gamma ct'\ ;\ \gamma\left(ct' + \dfrac{vx'}{c}\right)\ ;\ \text{none of these}\ \right]$. 49

Equating the above left and right hand sides gives

$$(x' + vt') = \left(ct' + \frac{vx'}{c}\right) \quad [\ true\ ;\ false\ ].$$ 80

$$x' - \left(\frac{v}{c}\right)x' = ct' - \left(\frac{v}{c}\right)ct'$$

$$x'\left(1 - \frac{v}{c}\right) = ct'\left(1 - \frac{v}{c}\right)$$

Hence $x' = ct'$ must be true if the equation $x = ct$ is true.

The purpose of this discussion is to give some idea of where the Lorentz transformation equations come from. It turns out that the Lorentz transformation equations are the only possible transformation that can make the speed of light the same in all reference systems; i.e., if the position of a light pulse is given by $x = ct$, then it is also given by $x' = ct'$ in a system that is moving with respect to the unprimed system.

Go to 11-8.

11-7
(from 11-13)

Inertial mass $M = W/c^2$ [*true* ; *false*]. 23

Go to 11-24.

11-8
(from 11-6)

Suppose observer A sees a boxcar of length $L = x_2 - x_1$ moving to the right at a velocity v. The primed system (x' and t') is "attached" to the boxcar and is used by observer B in all his measurements.

Observer B standing inside the boxcar measures the length of the boxcar to be $L' = x_2' - x_1'$. The values of x_1' and x_2' change with [t ; t' ; neither, they are constant]. 40

The length of the boxcar when at rest is [L ; L']. The length it appears to have observed in motion is [L ; L']. As measured in the reference system of [A ; B] the position x_2' at the end of the boxcar is related to x_2 by the equation $x_2' =$

[$\gamma(x_2 - vt)$; $\gamma(x_2 + vt)$; $x_2 - vt$; $x_2 + vt$; x_2] where $\gamma = \dfrac{1}{\sqrt{1 - (v/c)^2}}$. 89

11
76
53

Now write down the corresponding formula for x_1' and subtract to get $x_2' - x_1' =$

_____ . Hence $L =$ [$\gamma L'$; $\dfrac{L'}{\gamma}$; L'], which means that L, the measured length of the boxcar when [moving ; at rest] is [shorter ; longer] than L' by the factor $\sqrt{1 - (v/c)^2}$.

92
145
29
66

The moving car is [contracted ; expanded] in length. 81

The above calculation shows how the famous Lorentz contraction of moving lengths follows directly from the Lorentz transformation equation $x' = \gamma(x - vt)$.

Go to 11-9.

11-9
(from 11-8)

In this frame we shall derive the formula for time dilation. Observer A has a clock at rest fixed at $x = 0$. Observer B is traveling by to the right with velocity v. Then the time observed by [A ; B] is

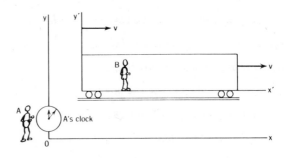

$t' = \gamma(t - vx/c^2)$. In this equation the position of the clock is given by $x =$ [vt ; $-vt$; $-\gamma vt$; 0 ; none of these] . Hence $t' = \left[\gamma t ; \dfrac{t}{\gamma} ; \dfrac{vt}{c} ; \text{none of these} \right]$.

If $t = 1$ min. A's clock reads [60 sec ; more than 60 sec ; less than 60 sec] , and t' is [greater ; less] than 60 sec. Hence clocks that read t' will be running [faster ; slower] than clocks that read t. Clocks of observer [A ; B] read t'. The clock in the figure reads [t ; t'] Hence as viewed from the primed system B's clocks run [faster ; slower] by the factor γ than the clock in the figure. In the figure, observer [A ; B] sees the moving clock. Hence moving clocks appear to be running [fast ; slow] when compared to clocks that are at rest with respect to the observer.

2
94
77
35
13
41
42
21
57
63
9

Go to 11-11.

11-10
(from 11-11)

If $v = 0.8c$, $\sqrt{1 - (v/c)^2} =$ _____ and $\gamma =$ _____ .

As measured from the earth, clocks on the rocket are [running fast by a factor of 1.67 ; running slow by a factor 0.6] . If the trip takes 25 years of earth time, it would take _____ times 25 years as measured by the [slower ; faster] clocks on the rocket.

A "meterstick" 10 light-years long when moving by with a velocity $v = 0.8c$ will appear [contracted ; expanded] to a length of _____ light-years.

12
47
22
31
60
93
75

Go to 11-14.

164

11-11
(from 11-9)

Suppose a superrocket could be built which could obtain a velocity of $v = 0.8c$. In one year this rocket could travel a distance of _____ light-years.

Traveling at this velocity, how long in earth time would it take for a round trip to a star at a distance of 10 light-years (as measured from the earth)? The answer is $t =$ _____ years. According to the clocks carried by the rocket, the round trip would take $t' =$ _____ years. The space traveler would age _____ years [more ; less] than his counterpart who remained on earth.

According to measurements from the rocket, the distance from the earth to the turn-around point was [5 ; 6 ; 8 ; 10 ; 12.5 ; 15] light-years.

102

141

73
84
62

27

Go to 11-10.

11-12
(from 11-19)

The formula $u' = \dfrac{u+v}{1+vu/c^2}$ holds true for all values of v whether positive or negative [*true* ; *false*] . If v is negative it means the [boxcar ; automobile] is moving to the [left ; right] . If we replace v in the formula with $(-v)$ we obtain the desired result:

$$u' = \frac{u + (-v)}{1 + (-v)u/c^2}.$$

We could also start from scratch using the Lorentz transformation between the stationary observer (primed system) and the boxcar moving to the left (unprimed system). These equations are $x' =$ [$\gamma(x - vt)$; $\gamma(x + vt)$] and $t' =$ [$\gamma(t - vx/c^2)$; $\gamma(t + vx/c^2)$] Dividing one by the other gives the correct answer.

5
68
43

55
14

Go to 11-17.

11-13
(from 11-28)

The value of inertial mass possessed by an amount of energy W is $M =$ [$\gamma W/c^2$; $W/\gamma c^2$; neither] .

6

If incorrect go to 11-7 ; if correct go to 11-24.

11-14
(from 11-10)

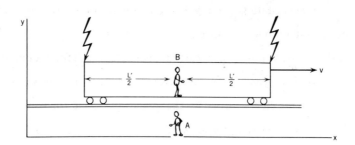

We see that, when moving at high velocities, time behaves in a "peculiar" way. Another one of these strange effects is that if two events occur simultaneously for observer A, the events will not in general be simultaneous for observer B.

Consider B standing in the center of a boxcar the length of which he measures to be L'. When the car comes to rest, the stationary observer A would measure the length to be $\left[\gamma L' \; ; \; \dfrac{L'}{\gamma} \; ; \; L' \right]$ where $\gamma = \dfrac{1}{\sqrt{1 - v^2/c^2}}$. Assume that when B passes A, two bolts of lightning strike the end of the boxcar simultaneously *according to A*. If the lightning bolts left marks on the track, A would measure the distance between marks to be $\left[\gamma L' \; ; \; \dfrac{L'}{\gamma} \; ; \; L' \right]$.

110

149

If incorrect go to 11-34 ; if correct go to 11-16.

11-15
(from 11-17)

Let us consider a neutron which decays into a proton, electron and antineutrino while traveling with a velocity of $v_n = 0.9c$. Assume the electron has $v_e = 0.8c$ with respect to the neutron; i.e., as would be measured by an observer moving along with neutron.

If v_n and v_e are parallel, the velocity of the electron as measured by a stationary observer would be _____ . If it is a backward decay; i.e., v_e and v_n are antiparallel, the electron velocity would be _____ .

124

100

Go to 11-22.

11-16
(from 11-14 or 11-34)

Referring back to frame 11-14, light from the two lightning bolts would reach A simultaneously [*true* ; *false*] . According to A, B is moving toward the right and [(a) both light pulses would reach B simultaneously ; (b) the pulse from the right would reach B first ; (c) the pulse from the left would reach B first] . **107**
161

According to B, both light pulses would reach him simultaneously [*true* ; *false*]. **139**

According to any observer the pulse from the right would reach B first [*true* ; *false*]. **144**

According to B, both pulses travel with velocity c along his boxcar. According to B, if the right pulse reaches him first, the right bolt of lightning must have struck first [*true* ; *false*]. **118**

We are forced to the conclusion that the same two events which occurred simultaneously according to one observer (A), did not occur simultaneously according to another observer (B).

Suppose a third observer had been moving to the left. The light pulse from the [left ; right] lightning bolt would reach him first and he would claim that the [left ; right] lightning bolt struck first. We see that the order in which these two events (lightning bolts striking) occur depends upon the velocity of the observer. Whenever two events occur within the time required for light to travel between them, the order of occurrence is undefined—it depends on the velocity of the observer. In such cases "future" events can be made to precede "past" events by selecting an appropriate moving observer. **113**
156

Go to 11-18.

11-17
(from 11-12)

Suppose the boxcar is traveling to the right, but the automobile has a velocity of u in the boxcar pointing to the left. Now $u' = \left[\dfrac{-u+v}{1-vu/c^2} \; ; \; \dfrac{-u+v}{1+vu/c^2} \; ; \; \dfrac{u-v}{1-vu/c^2} \; ; \; \dfrac{u+v}{1+vu/c^2} \; ; \; \text{none of these} \right]$ where v and u are positive numbers. **128**

Go to 11-15.

11-18

(from 11-16)

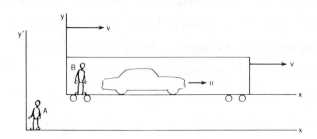

In this frame we will derive the Einstein addition of velocities formula. This time for convenience we use primed coordinates for the stationary observer A. Observer B is moving in a boxcar with velocity v in the unprimed system. B observes an automobile moving with velocity $u = \frac{x}{t}$ with respect to the boxcar (we define t to be 0 when $x = 0$).

The problem is: What value u' does A measure for the velocity of the auto? The relation between x' and x as given by the Lorentz transformation is $x' = \left[\; \gamma(x - vt) \;;\; \gamma(x + vt) \;;\; \frac{1}{\gamma}(x + vt) \;;\; \text{none of these} \;\right]$. Similarly $t' = \left[\; \gamma(t + vx/c^2) \;;\; \frac{1}{\gamma}(t + vx/c^2) \;;\; \text{neither} \;\right]$. If we divide the first equation by the second, we obtain

$$\frac{x'}{t'} = \frac{x + vt}{t + vx/c^2} \quad [\; true \;;\; false \;] \;.$$

111

123

96

If we divide numerator and denominator of the right side by t, we obtain $\frac{u + v}{1 + vu/c^2}$.

85

Since t' is zero when the automobile is at $x' = 0$, the left hand side is $[\; u \;;\; u' \;;\; u + u' \;]$. Hence, $u' = \frac{u + v}{1 + vu/c^2}$.

15

Go to 11-19.

11-19

(from 11-18)

Repeat the above derivation for the case when the boxcar is moving to the left with velocity v, and the automobile is still moving to the right in the boxcar with velocity u. The result is $u' =$ _____ where v and u are positive numbers.

26

If incorrect go to 11-12 ; if correct go to 11-20.

11-20
(from 11-19)
Let us consider a neutron which decays into a proton, electron, and antineutrino while traveling with a velocity of $v_n = 0.9c$. Assume the electron has $v_e = 0.8c$ with respect to the neutron; i.e., as would be measured by an observer moving along with the neutron. If v_n and v_e are parallel, the velocity of the electron as measured by a stationary observer would be _____ . If it is a backward decay, i.e., v_e and v_n are antiparallel, the electron velocity would be _____ .

3
25

If incorrect go to 11-22 ; if correct go to 11-21.

11-21
(from 11-20)
In this frame we shall show how the formula $M = \dfrac{M_0}{\sqrt{1-(v/c)^2}}$ is a mathematical consequence of the Lorentz transformation and the usual expression for conservation of momentum.

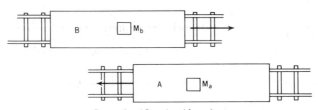

Boxcars A and B as viewed from above.

Consider two boxcars passing each other. Both boxcars contain identical masses; viz, when measured at rest they each have the value M_0. Just when the two cars are opposite each other, an attractive force between the two masses is momentarily turned on which gives a small change of momentum in the y-direction to each of the masses. Both observers will see their respective masses move in the y-direction with the same vertical component of

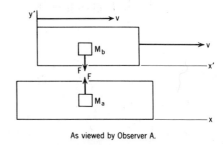

As viewed by Observer A.

11-21 (1) velocity [*true* ; *false*] . Hence $(v_a)_y$ = [$(v_b')_y$; $(v_b)_y$] where the prime
(continued) means as measured by B and no prime means as measured by A.

According to the formula for conservation of momentum the y-component of momentum given to M_a (as measured by A) must be equal to the y-component of momentum given to M_b (also as measured by A). The previous sentence says that

(2) $M_a(v_a)_y$ = [$M_b (v_b)_y$; $M_b' (v_b')_y$] . If we substitute the result of (1) into the [right- ; left-] hand side of (2) we obtain $M_a (v_b')_y = M_b (v_b)_y$

(3) $\qquad\qquad$ or $M_a \dfrac{\Delta y_b'}{\Delta t'} = M_b \dfrac{\Delta y_b}{\Delta t}$.

If M_b moves a distance $\Delta y_b'$, both observers measure the same distance; i.e., according to the Lorentz transformation $y' = y$ (the Lorentz contraction only affects distances in the direction of v) [*true* ; *false*].

Dividing both sides of (3) by Δy_b gives

(4) $\qquad \dfrac{M_a}{\Delta t'} = \dfrac{M_b}{\Delta t}$

However A observes that B's clocks are running [slower ; faster] than his. A observes that B measures [$\Delta t'$; Δt] to be [larger ; smaller] than the [$\Delta t'$; Δt] measured by A. The ratio $\dfrac{\Delta t}{\Delta t'}$ = $\left[\sqrt{1-(v/c)^2} \; ; \; \dfrac{1}{\sqrt{1-(v/c)^2}} \; ; \; 1 \right]$.

Substituting this into eq. 4 gives

M_b = _____ .

Since we are assuming v_y is small, M_a is essentially the rest mass M_0. Hence we have succeeded in showing that in order for the usual expression for conservation of momentum to hold true, the moving mass [M_a ; M_b] as observed by [A ; B] must have the value $\dfrac{M_0}{\sqrt{1-(v/c)^2}}$

69
83

17
4

50

87
99
112

20
44

71

91

58

Go to 11-22.

11-22
(from 11-15, 11-20, or 11-21)

The relativistic mass is $M = $ _____ M_0. 39

The relativistic momentum is $P = \left[Mv \; ; \; M_0 v \; ; \; \dfrac{Mv}{\sqrt{1 - v^2/c^2}} \right]$. 8

The relativistic total energy is $W = \left[Mc^2 \; ; \; M_0 c^2 \; ; \; (M - M_0)c^2 \right]$. 82

The relativistic kinetic energy is $\left[\dfrac{1}{2}Mv^2 \; ; \; \dfrac{1}{2}M_0 v^2 \; ; \; (M - M_0)c^2 \right]$. 114

Let $\beta = v/c$ and $\gamma = \dfrac{1}{\sqrt{1 - \beta^2}}$.

The ratio $Pc/W = \left[\gamma \; ; \; \gamma\beta \; ; \; \dfrac{\beta}{\gamma} \; ; \; \dfrac{\gamma}{\beta} \; ; \; \text{none of these} \right]$. 101

The ratio $P/M_0 c = \left[\gamma \; ; \; \gamma\beta \; ; \; \dfrac{\beta}{\gamma} \; ; \; \dfrac{\gamma}{\beta} \; ; \; \text{none of these} \right]$. 74

The ratio $W/M_0 c^2 = \left[\gamma \; ; \; \gamma\beta \; ; \; \dfrac{\beta}{\gamma} \; ; \; \dfrac{\gamma}{\beta} \; ; \; \gamma^2 \beta \; ; \; \text{none of these} \right]$. 30

If incorrect go to 11-26; if correct go to 11-23.

11-23
(from 11-22 or 11-26)

There is a simple relation giving the energy $W = Mc^2$ in terms of the momentum $P = Mv$ which we shall now derive.

$\gamma^2 (1 - \beta^2) = [\; 1 \; ; \; (1 - \beta^2)^2 \; ; \; (1 + \beta) \; ; \; \text{none of these} \;]$. 56

Hence $\gamma^2 = \gamma^2 \beta^2 + 1 \; [\; true \; ; \; false \;]$. 90

Now multiply both sides by $M_0^2 c^4$. Then

(1) $(M_0 \gamma c^2)^2 = (M_0 \gamma \beta c^2)^2 + (M_0 c^2)^2$.

The left-hand side is $[\; W \; ; \; W^2 \; ; \; (Pc)^2 \;]$. 86

The quantity $(M_0 \gamma \beta c^2)^2$ is $[\; W \; ; \; W^2 \; ; \; P^2 \; ; \; P^2 c^2 \;]$. 48

Substituting these into Eq. 1 gives $W^2 = P^2 c^2 + M_0^2 c^4 \; [\; true \; ; \; false \;]$. 127

Go to 11-28.

11-24
(from 11-13 or 11-7)

If the mass of a space traveler is 1.67 times his rest mass M_0, the kinetic energy of the space traveler will be $KE = \left[\frac{1}{2}M_0 v^2 \; ; \; \frac{1}{2}M_0(.8c)^2 \; ; \; .67 M_0 c^2 \; ; \; .67 Mc^2 \; ; \; .8 M_0 c^2 \; ; \; 1.67 M_0 c^2 \right]$.   98

The momentum of the space traveler will be $P = [\; M_0 v \; ; \; Mc \; ; \; 1.67 M_0 v \; ; \; 1.25 M_0 v \; ; \; \text{none of these} \;]$.   70
   37

The momentum will also be _____ $M_0 c$.

Go to 11-25.

11-25
(from 11-24)

Classically:

$KE = \left[\frac{1}{2}\beta^2 M_0 c^2 \; ; \; \frac{1}{2}\beta^2 \gamma^2 M_0 c^2 \; ; \; \text{neither} \right]$ where $\beta = \frac{v}{c}$ and $\gamma = \frac{1}{\sqrt{1-\beta^2}}$.   34

Classically, $P = \left[\gamma \; ; \; \beta \; ; \; \gamma\beta \; ; \; \frac{\beta}{\gamma} \; ; \; \gamma - 1 \right] M_0 c$.   7

Relativistically: $KE = [\; \gamma \; ; \; \gamma - 1 \; ; \; \gamma\beta \; ; \; \gamma\beta - 1 \;] M_0 c^2$.   116

$P = \left[\gamma \; ; \; \beta \; ; \; \gamma - 1 \; ; \; \gamma\beta \; ; \; \gamma\beta - 1 \; ; \; \frac{1}{\gamma} \; ; \; \frac{\beta}{\gamma} \right] M_0 c$.   150

$P = \left[\gamma \; ; \; \gamma - 1 \; ; \; \beta \; ; \; 1 \; ; \; \gamma\beta \; ; \; \frac{\beta}{\gamma} \; ; \; \frac{1}{\gamma} \right] Mc$.   135

Go to 11-27.

11-26
(from 11-22)

$Pc/W = (Mv)c/Mc^2$ [true ; false].   103

$Pc/W = v/c$ [true ; false].   108

$M = \left[M_0 \gamma \; ; \; \frac{M_0}{\gamma} \; ; \; M_0 \right]$.   65

$M/M_0 = \left[\gamma \; ; \; \frac{1}{\gamma} \; ; \; \gamma\beta \right]$.   54

$M/M_0 = W/M_0 c^2$ [true ; false].   143

Go to 11-23.

11-27
(from 11-25)

Let us assume that the universe is infinite in extent and contains a uniform density of stars separated by an average distance D. From this assumption we will deduce the striking conclusion that the entire night sky should be as bright as the sun.

The average number of stars in a cube of side D will be _____ .

32

Now draw a ray from the earth in an arbitrary direction out to infinity. As it passes through a cube of volume D^3 and cross sectional area D^2, the probability that it will hit a star of cross sectional area σ is one chance in $\left[\dfrac{1}{\sigma}\ ;\ \dfrac{D}{\sigma}\ ;\ \dfrac{D^2}{\sigma}\ ;\ \dfrac{D^3}{\sigma}\right]$. After the ray passes through $N = \left[\dfrac{D}{\sigma}\ ;\right.$ 140 / 115 $\left.\dfrac{D^2}{\sigma}\ ;\ \dfrac{D^4}{\sigma^2}\right]$ of these cubes in succession, its chances will be greater than 50-50 that it will hit a star. This means the average distance a ray must travel before hitting a star will be ND where $N = \left[\dfrac{\sigma}{D}\ ;\ \dfrac{D}{\sigma}\ ;\ \dfrac{D^2}{\sigma}\right]$.

153

Astronomical measurements show that the average distance between stars is $D = 10$ lightyears and that the average area of a star is $\sigma \approx 10^{-14}$ light-years^2. Hence $\dfrac{D^3}{\sigma} = \dfrac{10^3}{10^{-14}} = 10^{17}$ light-years. We are forced to the conclusion that if the universe is larger than _____ light-years and if the stars are uniformly distributed, with the same surface temperature or brightness, then all rays from the earth will end in stars having the same brightness as the sun; i.e., all portions of the sky would be as bright as the sun; i.e., all portions of the sky would be as bright as the sun. In such a situation the temperature of the earth would quickly become the same as the surface temperature of the sun [*true* ; *false*].

159

131

Go to 11-29.

11-28
(from 11-23)

Every form of energy has an inertial mass associated with it [*true* ; *false*].

120

Go to 11-13

11-29
(from 11-27)
Suppose there was a blob of interstellar fog in such a universe. The temperature of the fog would become the same as the surface temperature of the sun [*true* ; *false*]. 163
The presence of interstellar fog could explain why the sky is dark [*true* ; *false*]. 133
From the above logical reasoning we must conclude that either the universe is finite in size or else the density of stars decreases at large distances [*true* ; *false*]. 105

Go to 11-30.

11-30
(from 11-29)
There is another way out of this paradox (called Olbers' paradox). Space could still be infinite and populated everywhere with a uniform density of stars if the distant stars are rapidly moving [toward ; away from] the earth. The wavelength of the light 106
emitted by a star moving near the speed of light [toward ; away from] the earth 147
will be "red shifted" from the visible to much longer wavelengths. This red shift of distant galaxies is observed [*true* *false*] . 125

Go to 11-32.

11-31
(from 11-32)
The steady state theory of cosmology assumes that the generalized principle of uniformity is true [*true* ; *false*]. 36
If the steady state theory is true, then $n = 1$ in the velocity-distance relation $v = Kr^n$ [*true* ; *false*]. 64
If astronomical observations should show that n is not equal to one, then the steady state theory would be disproven [*true* ; *false*]. 109
The Hubble constant K has units of [sec/cm ; sec ; 1/sec ; cm sec] . 97
The generalized principle of uniformity states that the universe does not change with time or position of the observer [*true* ; *false*]. 119

Go to 11-33.

11-32
(from 11-30)

The rather meager experimental information indicates that the velocity of [recession ; approach] of distant galaxies is proportional to the [square root ; first ; 2nd ; 3rd] power of the distance. This velocity-distance relationship can be expressed as $v = Kr^n$ where K is a universal physical constant (called Hubble's constant) and the exponent $n =$ _____ .

129
166
158

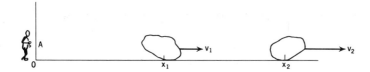

We will now show that the principle of uniformity predicts what n should be. Let us consider two nearby and slowly-moving galaxies with velocities v_1 and v_2. The velocity of galaxy 1 as observed by observer A would be $v_1 =$ _____ .

134

Observer A measures the relative velocity between galaxies 1 and 2 to be $(v_2 - v_1)$ which is [$K(x_2 - x_1)^n$; $K(x_2{}^n - x_1{}^n)$; $(Kx_2 - Kx_1)^n$] . According to the principle of uniformity an observer on galaxy 1 would see galaxies receding from him according to the law $v = Kr^n$ where [(a) r is measured from the origin $(x = 0)$; (b) r is the distance from galaxy 1 (measured from $x = x_1$)] . The observer on galaxy 1 would say that the velocity of galaxy 2 is [$Kx_2{}^n$; $K(x_2 - x_1)^n$; $K(x_2{}^n - x_1{}^n)$] . If we equate the two above results for the velocity of 2 with respect to 1 we obtain $(x_2 - x_1)^n = x_2{}^n - x_1{}^n$.

117
142
152

According to the principle of uniformity, for velocities much smaller than c the above equation must be true for all values of x_1 and x_2 [*true* ; *false*]. For example, take $x_2 = 3$ million light-years and $x_1 = 1$ million light-years. Then $(3-1)^n = 3^n - 1^n$. The above equation is true when $n =$ _____ .

121
160

Go to 11-31.

11-33
(from 11-31)
Suppose the matter in the universe were distributed uniformly with a density D out to a distance R. The total mass M of the universe would be $\left[\dfrac{D}{\frac{4}{3}\pi R^3} \; ; \; \dfrac{4}{3}\pi R^3 D \; ; \; 4\pi R^2 D \right]$. In order for this "universe" not to collapse the galaxies at the outer edge must have the velocity of escape from a sphere of mass M and radius R. Hence the kinetic energy of such a galaxy must equal _____ times the gravitational potential energy, or $\frac{1}{2}mv^2 = GMm/R$ where m is the mass of the [galaxy ; universe] .

167

67

130

Hence $v = \sqrt{2GM/R}$ is the velocity of escape.

Now substitute into this the expression for M in terms of D:

then $v = $ _____ .

148

Note that this is the same form as the Hubble relationship for the velocity of recession of distant galaxies which is $v = KR$. What is even more surprising is that if we put in the measured values of G and D, we find that $\sqrt{\dfrac{8}{3}\pi GD} = K$ within experimental errors. Actually within the errors $GD = K^2$. This may be pure coincidence. Or perhaps, there is a deep, theoretical meaning to it which we at present do not understand. It may be that the universal physical constants G, D (density of matter in the universe), and K (the Hubble constant) are not independent.

Go to 11-35.

11-34
(from 11-14)
When A measures the length of the moving car he will find that its length is [contracted ; expanded ; the same] as when at rest. Observer A can use the stationary train track as a "meter stick" to measure the car length if the position of the two ends of the car are observed simultaneously in the reference system of [A ; B] . The lightning bolt marks on the track can serve as such a measurement [true ; false] .

168

162

151

Go to 11-16.

11-35
(from 11-33)

The relation $GD \approx K^2$ can be given another theoretical interpretation. If there is a spontaneous creation of matter (as required by the [steady state ; big bang] theory), we would like the law of conservation of energy not to be violated. If a mass m is spontaneously created the total increase in energy of the universe should be $\Delta W = mc^2 + U$ where U is the gravitational potential energy of m. $U =$ [$-GM/R$; $-GMm/R$; $-GMm/R^2$] where M is the total mass and R the average distance of the "visible" galaxies (there is a "red shift" of the gravitational force as well as of light). Then $M \approx DR^3$ where D is the average density of matter in the universe. Hence $U \approx$ _____. The increase in energy becomes $\Delta W \approx mc^2 - GDmR^2$. The radius of the "visible" universe is $R \approx$ _____ where K is the Hubble constant. Then $\Delta W \approx m(c^2 - GDc^2/K^2) = mc^2(1 - GD/K^2)$ [true ; false]. If $GD = K^2$, the above energy increase becomes zero and at least from this point of view energy conservation is not violated by spontaneous creation.

165

137

122
155
132

Go to 11-36.

11-36
(from 11-35)

Review Questions:

Express the following quantities in terms of β, γ, M_0 and c where $\beta = \dfrac{v}{c}$ and $\gamma = \dfrac{1}{\sqrt{1-\beta^2}}$

$P =$ _____ 136

$W =$ _____ 126

$P/W =$ _____ 157

$W^2 - c^2 P^2 =$ _____ 169

$KE/W =$ _____ 146

$\sqrt{KE(KE + 2M_0 c^2)} =$ _____ 154

Express $\dfrac{(\gamma^2 - 1)}{\gamma^2}$ in terms of β only. $\dfrac{\gamma^2 - 1}{\gamma^2} =$ _____ 164

177

Chapter 11 Answer Sheet

Item Number	Answer	Item Number	Answer
1	$10\sqrt{3}$	46	vt
2	B	47	1.67
3	$\dfrac{1.7c}{1.72}$	48	$p^2 c^2$
4	left	49	$\gamma\left(ct' + \dfrac{vx'}{c}\right)$
5	true	50	true
6	neither	51	true
7	β	52	$-vt'$
8	Mv	53	B
9	slow	54	γ
10	$\gamma(x' + vt')$	55	$\gamma(x - vt)$
11	L'	56	1
12	0.6	57	faster
13	greater	58	A
14	$\gamma\left(t - \dfrac{vx}{c^2}\right)$	59	more
15	u'	60	slower
16	shorter	61	greater
17	$M_b (v_b)_y$	62	less
18	6	63	B
19	ct'	64	true
20	Δt	65	$M_0 \gamma$
21	t	66	shorter
22	running slow by a factor of 0.6	67	1
23	true	68	boxcar
24	2	69	true
25	$\dfrac{0.1c}{0.28}$	70	$1.67 M_0 v$
26	$\dfrac{u - v}{1 - vu/c^2}$	71	$\dfrac{M_a}{\sqrt{1 - (v/c)^2}}$
27	6	72	6
28	7	73	15
29	moving	74	$\gamma \beta$
30	γ	75	6
31	0.6	76	L
32	1	77	γt
33	greater	78	less
34	$\dfrac{1}{2}\beta^2 M_0 c^2$	79	2
35	60 sec	80	true
36	true	81	contracted
37	1.33	82	Mc^2
38	more	83	$(v_b')_y$
39	$\left(1 - \dfrac{v^2}{c^2}\right)^{-\frac{1}{2}}$	84	10
40	neither	85	$\dfrac{u + v}{1 + vu/c^2}$
41	faster	86	W^2
42	B	87	slower
43	left	88	false
44	$\dfrac{1}{\sqrt{1 - (v/c)^2}}$	89	$\gamma(x_2 - vt)$
		90	true
		91	M_b
		92	$\gamma(x_2 - x_1)$
45	true	93	contracted
		94	zero
		95	true

Item Number	Answer
96	true
97	1/sec
98	$0.67 M_0 c^2$
99	$\Delta t'$
100	$\dfrac{0.1c}{0.28}$
101	none
102	0.8
103	true
104	same as
105	false
106	away from
107	true
108	true
109	true
110	L'
111	$\gamma(x + vt)$
112	smaller
113	left
114	$(M - M_0)c^2$
115	$\dfrac{D^2}{\sigma}$
116	$\gamma - 1$
117	$K(x_2^n - x_1^n)$
118	true
119	true
120	true
121	true
122	$-GDmR^2$
123	$\gamma\left(t + \dfrac{vx}{c^2}\right)$
124	$\dfrac{1.7c}{1.72}$
125	true
126	$\gamma M_0 c^2$
127	true
128	$\dfrac{-u + v}{1 - vu/c^2}$
129	recession
130	galaxy
131	true
132	true
133	false
134	Kx_1^n
135	β
136	$\gamma\beta M_0 c$

Item Number	Answer
137	$\dfrac{-GMm}{R}$
138	true
139	false
140	$\dfrac{D^2}{\sigma}$
141	25
142	(b)
143	true
144	true
145	$\dfrac{L'}{\gamma}$
146	$1 - \dfrac{1}{\gamma}$
147	away from
148	$R\sqrt{\dfrac{8}{3}\pi GD}$
149	$\dfrac{L'}{\gamma}$
150	$\gamma\beta$
151	true
152	$K(x_2 - x_1)^n$
153	$\dfrac{D^2}{\sigma}$
154	$\gamma\beta M_0 c^2$
155	$\dfrac{c}{K}$
156	left
157	$\dfrac{\beta}{c}$
158	1
159	10^{17}
160	1
161	(b)
162	A
163	true
164	β^2
165	steady state
166	first
167	$\dfrac{4}{3}\pi R^3 D$
168	contracted
169	$M_0^2 c^4$

Chapter 12
QUANTUM THEORY

12-1 Rutherford determined the mass of the atomic nucleus by the [number ; energy] **12**
of [alpha particles ; electrons] that were scattered backwards. The scattering **35**
mechanism is the [gravitational ; electrostatic ; magnetic ; nuclear] force. **19**

Go to 12-2.

12-2 Consider a box of cross sectional area A.
(from 12-1) The box contains three apples each with
a cross sectional area σ. The total cross
sectional area presented by the apples is
_____ . An arrow aimed at the box **18**
would have a chance of _____ of hitting an apple. If N_a arrows are shot into the box, **76**
the number hitting apples would be _____ . **23**

Go to 12-3.

12-3 In the next two frames we will see how the charge of the nucleus is determined by
(from 12-2) counting the number of scattered alpha particles.
Suppose each nucleus has a cross sectional area of σ cm^2. Suppose one alpha particle is
shot at a metal foil of cross sectional area A containing N nuclei. The chance of hitting
one of the nuclei would be the ratio of the cross sectional areas or _____ . If the **40**
total number of alpha particles shot at the foil is N_a, the number of scattered alphas
would be N_a times _____ . **33**

Go to 12-6.

181

12-4　If n is the number of alphas scattered more than 90°, then the experimental determina-　　**49**
(from　tion of σ is given by $\sigma =$ _____ where N is the number of [electrons ; atoms]　　**54**
12-6 or
12-12)　in the foil. A is the total area of all the nuclei [*true ; false*] .　　**13**

　　If you are confused where this formula came from, solve for σ in Frame 12-3.

　　Go to 12-7.

12-5　The ratio of electron to proton mass is about _____ . In an atom such as carbon the　　**7**
(from　fraction of atomic mass not in the nucleus is about _____ . This is the ratio of masses　　**22**
12-7)
　　of 6 electrons to _____ nucleons (neutrons or protons).　　**68**

　　Go to 12-8.

12-6
(from 12-3)

In this frame we shall calculate the effective cross sectional area σ of a nucleus. Let Q be the charge of the nucleus and q the charge of the alpha particle. The distance b from the initial line of flight is called the impact parameter. The distance of closest approach will be [b ; greater than b ; less than b] .   41

The angle of deflection θ [increases ; decreases] with increasing b. The exact relation between impact parameter and θ can be calculated using [Coulomb's ; Ampere's ; Faraday's] law $\left(\text{the result of such a calculation turns out to be } b = \dfrac{Qq}{mv^2 \tan\frac{\theta}{2}}\right)$. For a deflection of 90°, $b = $ _____ . If the initial direction of the alpha is such that b is less than Qq/mv^2, then it will be deflected [more ; less] than 90°. This corresponds to a region around the nucleus of area σ where $\sigma = \left[\, \pi\left(\dfrac{Qq}{mv^2}\right)^2 \;;\; 2\pi\left(\dfrac{Qq}{mv^2}\right)^2 \;;\; 4\pi\dfrac{Qq}{mv^2} \,\right]$.   48, 26, 81, 61, 72

Hence $\sigma = \left[\, \dfrac{\pi Q^2 q^2}{(\text{KE})^2} \;;\; \dfrac{\frac{1}{4} Q^2 q^2}{(\text{KE})^2} \;;\; \dfrac{\frac{1}{4}\pi Q^2 q^2}{(\text{KE})^2} \;;\; \dfrac{\frac{1}{2}\pi Q^2 q^2}{(\text{KE})^2} \,\right]$ is the effective cross sectional area for scattering alphas [at any angle ; more than 90° ; less than 90°] .   37, 92

For N_a alphas hitting a foil of area A, the number alphas scattered more than 90° would be $\left[\, \dfrac{N_a N \sigma}{A} \;;\; \dfrac{N_a N \sigma^2}{A} \,\right]$ where N is the number of [alphas hitting the foil ; alphas scattered more than 90° ; atoms in the foil] .   86, 20

Thus we see that by measuring the number of alphas scattered more than 90° the effective cross sectional area σ can be experimentally determined. Then the charge of the nucleus can be determined from the formula $Q = \left[\, \dfrac{2\text{KE}}{q}(\sigma/\pi) \;;\; \dfrac{2\text{KE}}{\pi q}\sqrt{\sigma} \;;\; \dfrac{2\text{KE}}{q}\sqrt{\dfrac{\sigma}{\pi}} \,\right]$.   53

If incorrect go to 12-12 ; if correct go to 12-4.

12-7
(from 12-4)
Rutherford found that the charge of the nucleus was $Q = \left[Z \; ; \; Ze \; ; \; \frac{1}{2}Ze \right]$ where 67

Z is the atomic number and e is the charge of the electron. He also found that all but

$\left[\frac{1}{2} \; ; \; \frac{1}{4} \; ; \; \frac{1}{Z} \; ; \; \text{none of these} \right]$ of the mass of the atom was in the nucleus. 44

If incorrect go to 12-5 ; if correct go to 12-8.

12-8
(from 12-5 or 12-7)
Let W be the energy of a photon of frequency f and wavelength λ.

Then $f = \frac{h}{W}$ [*true ; false*] . $\lambda = \frac{h}{W}$ [*true ; false*] .   17
                                                                                          9

$\lambda = \frac{c}{f}$ [*true ; false*] . $\lambda = \frac{hc}{W}$ [*true ; false*] .   30
                                                                                              98

Go to 12-18.

12-9
(from 12-18)
A 5 ev photon is absorbed by an electron on the surface of a metal. The work function of this metal is $w = 2$ ev. The kinetic energy of the electron after leaving the metal will be
_____ ev.   101

Go to 12-10.

12-10
(from 12-9)
Suppose the previous electron loses 1 ev by collisions inside the metal before it leaves the surface. Then the kinetic energy of this electron after leaving the metal would be
_____ ev.   3

Go to 12-14.

12-11
(from 12-19)
In this closed system of electron plus metal, mechanical energy will be conserved (assuming no friction) [*true ; false*] .   5

Until this electron collides with other electrons the quantity $KE + U$ will be zero whether it is inside or outside the metal. Inside the metal $KE = 5$ ev and $U =$ _____ ev.   39

Go to 12-20.

12-12	We have already derived the result $\sigma = \dfrac{\pi Q^2 q^2}{4(KE)^2}$ [*true* ; *false*] .	64
(from 12-6)	This equation can be solved for Q^2 by multiplying both sides by _____ . The result is	78
	$Q^2 = \dfrac{4(KE)^2 \, \sigma}{\pi q^2}$ [*true* ; *false*] .	106
	σ is determined experimentally by measuring the [energy ; number] of the alphas scattered more than 90°.	94

Go to 12-4.

12-13 (from 12-31) As the energy of a photon decreases, its wavelength [increases ; decreases ; remains the same] . 　　1

If incorrect go to 12-23 ; if correct go to 12-24.

12-14 (from 12-10) According to the theory of metals conduction electrons have kinetic energy while in the metal and are bound to the metal by an attractive electrostatic force. For a certain metal the maximum kinetic energy of the conduction electrons in the metal is 3 ev, but these electrons would need an additional 2 ev to leave the surface. The work function of this metal is _____ ev. Electrons having a total kinetic energy of _____ ev would just barely be able to leave this metal and would lose _____ ev of kinetic energy in the process of leaving the metal. Suppose one of these 3 ev electrons absorbs a 4 ev photon. The kinetic energy of this electron will be _____ ev before leaving the surface. After leaving the surface its kinetic energy will be _____ ev less or KE_{final} = _____ ev.

38
11
50

43
62
91

If incorrect go to 12-15 ; if correct go to 12-16.

12-15 (from 12-14) In the previous problem a surface electron absorbs a photon of energy hf = 4 ev. If the work function w = 2 ev, the kinetic energy left over after leaving the metal will be [$w - hf$; $hf - w$; $hf + w$] which is _____ ev.

109
97

Go to 12-16.

185

12-16
(from 12-14 or 12-15)
In leaving the surface of a metal the kinetic energy of an electron is decreased by an amount equal to the work function [*true* ; *false*] .

96

Go to 12-17.

12-17
(from 12-16)
The additional kinetic energy necessary for an electron to leave the metal is the work function [*true* ; *false*] .

55

Go to 12-19.

12-18
(from 12-8)
$hf = W$ [*true* ; *false*] .

71

$h(c/\lambda) = W$ [*true* ; *false*] . Solving for λ gives $\lambda = $ _____ .

46
28

Go to 12-9.

12-19
(from 12-17)
An electron outside a metal surface has KE = 0 and potential energy $U = 0$. Its total energy (KE + U) is zero [*true* ; *false*] . The positive atomic nuclei of the metal will attract this electron [*true* ; *false*] . Suppose this attractive potential gives the electron 5 ev of kinetic energy by the time it has entered the metal. The sum (KE + U) inside the metal will be _____ ev. If (KE + U) = 0 and KE = 5 ev, then $U = $ _____ ev.

88
51

21
32

Go to 12-11.

12-20
(from 12-11)
Inside this metal $U = -5$ ev. If the maximum KE of the conduction electrons is 3 ev, it will take _____ ev of additional energy to remove one of these conduction electrons from the surface. Hence the work function would be _____ ev.
We see that there is a simple relation between work function w, potential energy U, and the maximum kinetic energy KE_0 of the conduction electrons. This relation is $-U = $ [$KE_0 + w$; $KE_0 - w$; $w - KE_0$] where U is a [positive ; negative] number. This relation will be studied in more detail in Chapter 14.

2
47

15
58

Go to 12-21.

12-21
(from 12-20)
In the photoelectric effect the maximum kinetic energy of the emitted electrons is related to the work function by the equation $w =$ [$hf - KE_{max}$; $KE_{max} - hf$; KE_{max}] .   90

Go to 12-22.

12-22
(from 12-21)
In the photoelectric effect a plot of the maximum KE of the electrons vs. the photon energy W would look like curve [1 ; 2 ; 3] .   97

In the above curve, the work function w has the value $\left[a \ ; \ b \ ; \ \dfrac{1}{b} \ ; \ \text{the slope} \right]$.   102

Go to 12-31.

12-23
(from 12-13)
For the solution of the previous problem refer to Frame 12-18.

Go to 12-24.

12-24
(from 12-13 or 12-23)
A plot of λ vs. W for a photon would look like curve _____ .   25

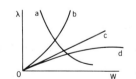

Go to 12-25.

12-25
(from 12-24)
According to relativity theory, inertial mass is always given by

$M = \left[\dfrac{W_0}{c^2} \ ; \ \dfrac{KE}{c^2} \ ; \ \dfrac{W}{c^2} \right]$. The mass of a 10 ev (1.6×10^{-11} ergs) photon is   57

$M =$ _____ grams.   83

Go to 12-36.

12-26
(from 12-37)
The numerical relation between λ and KE of a photon is KE(in ev) = $\frac{12390}{\lambda(\text{in }A)}$

[*true* ; *false*] . 66

Go to 12-27.

12-27
(from 12-26)
The numerical relation between λ and KE of an electron is λ (in A) = $\frac{12390}{\text{KE (in ev)}}$

[*true* ; *false*] . 100

Go to 12-29.

12-28
(from 12-29)
For a photon λ = $\left[\ \dfrac{hc}{W}\ ;\ \dfrac{hW}{c}\ ;\ \dfrac{hc^2}{W}\ \right]$ 104

The kinetic energy of a photon is the same as its total energy W [*true* ; *false*] . 74

Hence λ = hc/KE for [photons ; electrons ; both photons and electrons] . 60

Go to 12-32.

12-29
(from 12-27)
For small velocities, the relation between KE and momentum is KE = $\left[\ \dfrac{mP^2}{2}\ ;\ \dfrac{P^2}{2m}\ ;\right.$ 16

$\left.\dfrac{P^2}{2m^2}\ \right]$. The momentum of a non-relativistic electron in terms of its kinetic energy is 70

P = _____ . The wavelength of an electron in terms of its KE is λ = h/P [*true* ; 80

false] . The wavelength of an electron in terms of its KE is λ _____ . 45

This equation also holds true for a photon [*true* ; *false*] . 27

Go to 12-28.

12-30
(from 12-34)
If the amplitude is 3 times as much, the wave intensity at electron A is _____ times as much as at B. The number of photons passing by A is _____ times as many as are passing by B. 6

24

Go to 12-35.

12-31
(from 12-22)

$KE_{max} = [\ W-w\ ;\ w-W\ ]$ where $W = hf$.

When $KE_{max} = 0$, $W = w$ [*true* ; *false*] .

Point [*A* ; *B*] corresponds to $KE_{max} = 0$.

The value of this point as read off the [vertical ; horizontal] axis will be w.

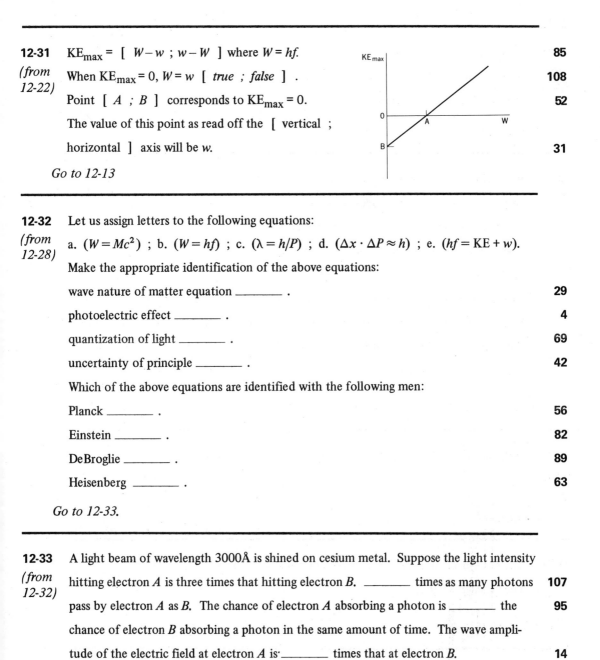

Go to 12-13

85
108
52

31

12-32
(from 12-28)

Let us assign letters to the following equations:

a. $(W = Mc^2)$; b. $(W = hf)$; c. $(\lambda = h/P)$; d. $(\Delta x \cdot \Delta P \approx h)$; e. $(hf = KE + w)$.

Make the appropriate identification of the above equations:

wave nature of matter equation _____ .   29

photoelectric effect _____ .   4

quantization of light _____ .   69

uncertainty of principle _____ .   42

Which of the above equations are identified with the following men:

Planck _____ .   56

Einstein _____ .   82

DeBroglie _____ .   89

Heisenberg _____ .   63

Go to 12-33.

12-33
(from 12-32)

A light beam of wavelength 3000Å is shined on cesium metal. Suppose the light intensity hitting electron *A* is three times that hitting electron *B*. _____ times as many photons pass by electron *A* as *B*. The chance of electron *A* absorbing a photon is _____ the chance of electron *B* absorbing a photon in the same amount of time. The wave amplitude of the electric field at electron *A* is _____ times that at electron *B*.

107
95

14

Go to 12-34.

12-34
(from 12-33)
If the amplitude of the light wave shining on electron A is 3 times that shining on electron B, the chance of electron A absorbing a photon is [1 ; $\sqrt{3}$; 3 ; 9] times that of electron B absorbing a photon.

75

If incorrect go to 12-30 ; if correct go to 12-35.

12-35
(from 12-34 or 12-30)
Blue light of wavelength 4000Å strikes a double slit. At point P_1 the wave amplitudes from slits A and B are 10 and –9 units respectively. Hence point P_1 is near an interference [maximum ; minimum ; neither] .

8

The resultant wave amplitude at P_1 would be [1 ; –1 ; 19 ; none of these] units.

34

If the wave amplitudes from slits A and B are +9 and +10 units when at P_2, the ratio of the amplitude at P_2 to that at P_1 is _____ . The number of photons per second passing P_2 is _____ times the number of photons per second passing P_1.

59

77

If incorrect go to 12-39 ; if correct go to 12-38.

12-36
(from 12-25)
If the mass of a photon is $M = hf/c^2$, its momentum will be $P = $ _____ .

10

Go to 12-37.

12-37
(from 12-36)
The momentum of a photon is its velocity times (hf/c^2) [true ; false] .

99

The velocity is v = _____ .

87

Go to 12-26.

12-38
(from 12-35 or 12-39)
In the ground state of the hydrogen atom the wave amplitude of the electron is maximum at the position of the proton and drops off exponentially with distance from the center. At a distance of 0.5×10^{-8} cm the wave amplitude is 1/2.7 times its value at the center. The chance of finding the electron at the center of the hydrogen atom is _____ times the chance of finding it at a point 0.5×10^{-8} cm from the center.

103

Go to 12-40.

12-39 If the ratio of the amplitudes is 19, the ratio of the intensities will be _____ . The   **105**
(from number of photons per sec is proportional to the wave [amplitude ; intensity] .   **79**
12-35)

Go to *12-38*.

12-40 Review Questions:
(from
12-38) 1. What is the energy of a photon in terms of *h, c,* and *f*?

$$W = \underline{\hspace{1cm}}$$ **36**

2. What is the kinetic energy of a non-relativistic electron in terms of *m, h,* and λ?   KE = _____   **84**

3. The probability of finding a particle is proportional to the wave amplitude. True or false? _____   **73**

4. If the work function of a metal is 4 ev, what will be the minimum photon energy capable of producing photoelectrons from this metal?

$$W = \underline{\hspace{1cm}} \text{ ev.}$$ **65**

What will be the wavelength of such photons?

$$\lambda = \underline{\hspace{1cm}}$$ **95**

Chapter 12 Answer Sheet

Item Number	Answer	Item Number	Answer	Item Number	Answer
1	increases	45	$\dfrac{h}{\sqrt{2mKE}}$	87	c
2	2	46	true	88	true
3	2	47	2	89	c
4	e	48	decreases	90	$hf - KE_{max}$
5	true	49	$\dfrac{nA}{N_a N}$	91	2
6	9	50	5	92	more than 90°
7	$\dfrac{1}{2000}$	51	true	93	3 times
8	minimum	52	A	94	number
9	false	53	$\dfrac{2KE}{q}\sqrt{\dfrac{\sigma}{\pi}}$	95	3085Å
10	$\dfrac{hf}{c}$	54	atoms	96	false
11	5	55	true	97	2
12	energy	56	b	98	true
13	false	57	$\dfrac{W}{c^2}$	99	true
14	$\sqrt{3}$	58	negative	100	false
15	$KE_0 + w$	59	19	101	3
16	$\dfrac{P^2}{2m}$	60	photons	102	b
17	false	61	more	103	2.7^2
18	3σ	62	5	104	$\dfrac{hc}{W}$
19	electrostatic	63	d	105	19^2
20	atoms in the foil	64	true	106	true
21	zero	65	4	107	3
22	$\dfrac{1}{4000}$	66	true	108	true
23	$N_a \dfrac{3\sigma}{A}$	67	Ze	109	$hf - w$
24	9	68	12		
25	a	69	b		
26	Coulomb's	70	$\sqrt{2mKE}$		
27	false	71	true		
28	$\dfrac{hc}{W}$	72	$\pi\left(\dfrac{Qq}{mv^2}\right)^2$		
29	c	73	false		
30	true	74	true		
31	horizontal	75	9		
32	-5	76	$\dfrac{3\sigma}{A}$		
33	$\dfrac{N\sigma}{A}$	77	361		
34	1	78	$\dfrac{4(KE)^2}{\pi q^2}$		
35	alpha particles	79	intensity		
36	hf	80	true		
37	$\dfrac{\frac{1}{4}\pi Q^2 q^2}{(KE)^2}$	81	$\dfrac{Qq}{mv^2}$		
38	2	82	a, e		
39	-5	83	1.8×10^{-32}		
40	$\dfrac{N\sigma}{A}$	84	$\dfrac{h^2}{2m\lambda^2}$		
41	greater than b	85	$W - w$		
42	d	86	$\dfrac{N_a N\sigma}{A}$		
43	7				
44	none				

Chapter 13
ATOMIC THEORY

13-1 In this frame we will show that the resolution of a microscope is limited by the uncertainty principle. With a microscope, the position of a particle can be located to within about one wavelength of the light being used. Hence the uncertainty in [position ; momentum ; velocity] is $\Delta x = \left[\lambda \; ; \dfrac{1}{\lambda} \right]$.

22
102

In order to detect the particle at least one of the photons in the converging beam of light must either be scattered or absorbed by the particle. Hence the momentum given to the particle might be as much as $\left[hf \; ; \dfrac{hf}{c} \; ; \dfrac{hf}{c^2} \right]$ which also has the value $\left[\dfrac{h}{\lambda} \; ; h\lambda \; ; \dfrac{h}{\lambda^2} \; ; \dfrac{hc}{\lambda} \right]$. The uncertainty in momentum of the particle will be of this order of magnitude [true ; false]. Hence $\Delta P \approx \left[h\lambda \; ; \dfrac{h\lambda}{c} \; ; \dfrac{h}{\lambda} \right]$.

40
73

56
93

The product of the above expression for Δx times the above result for ΔP is (λ) times (h/λ) or h. We see that our result agrees with the uncertainty principle which states: $(\Delta x)(\Delta P) \approx h$.

Go to 13-3.

13-2 The lowest standing wave has _____ wavelengths from one end of the box to the other which distance is $\left[\dfrac{1}{2}R \; ; R \; ; 2R \right]$. Hence $\dfrac{1}{2}\lambda = 2R$ [true ; false].

(from 13-3)

81
47
12

Go to 13-5.

13-3
(from 13-1)

The curve to the right represents the potential energy of a box of length _____ . If an electron has a total energy W_0 its kinetic energy will be [a ; b ; $(a+b)$] . The magnitude of its potential energy inside the box will be [a ; b ; $(a+b)$] .

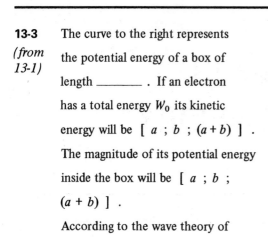

According to the wave theory of matter the electron must have a wavelength $\lambda = \left[hp \; ; \; \dfrac{h}{p} \; ; \; \dfrac{p}{h} \right]$ where p is the momentum.

In terms of its kinetic energy the wavelength must be $\lambda = $ _____ .

The lowest possible standing wave in the box has $\lambda_1 = \left[\dfrac{1}{2}R \; ; \; R \; ; \; 2R \; ; \; 4R \right]$.

1
69
75
53
29
17

If incorrect go to 13-2 ; if correct go to 13-5.

13-4
(from 13-14)

The $N = 2$ state can have $\ell = 3$ [*true* ; *false*] . 77
The $N = 2$ state can have $\ell = 2$ [*true* ; *false*] . 31
The $N = 2$ state can have $\ell = 1$ and 0 [*true* ; *false*] . 62
The $\ell = 0$ wave can hold _____ electrons. 67
The $\ell = 1$ wave can hold _____ electrons. 11
The sum of the above number of electrons is _____ . 26

Go to 13-15.

13-5
(from 13-2 or 13-3)

The Nth order standing wave in the above box would have a wavelength $\lambda_N = \left[\dfrac{R}{2N} \; ; \; \dfrac{R}{N} \; ; \; \dfrac{2R}{N} \; ; \; \dfrac{4R}{N} \right]$.

4

If incorrect go to 13-7 ; if correct go to 13-6.

13-6
(from 13-5 or 13-7)

Consider an electron in a circular orbit around a nucleus of charge Ze. The kinetic energy will be _____ times the magnitude of the potential energy. This is easily seen by [(a) using conservation of energy ; (b) using conservation of momentum; (c) equating centripetal force with electrostatic force] .

Then $\dfrac{mv^2}{R} = \dfrac{Ze^2}{R^2}$.

Now multiply both sides by $\dfrac{R}{2}$. Then $\dfrac{1}{2}mv^2 = \dfrac{1}{2}Ze^2/R$ which is _____ times U, the potential energy.

30

6

44

Go to 13-10.

13-7
(from 13-5)

For the Nth order standing wave the number of half wavelengths in a length L is _____. The number of wavelengths would be _____ . The wavelength would be L divided by this or $\lambda_N = $ _____ . L has the value $\left[\dfrac{1}{2} \; ; \; 1 \; ; \; 2\right]$ times R.

85

36
100
70

Go to 13-6.

13-8
(from 13-13)

The Pauli principle tells us that for each possible electron orbital wave function there can be no more than _____ electron(s).

9

Go to 13-14.

13-9
(from 13-10)

In this frame we shall go through an approximate derivation of the energies of the hydrogen atom using the modern wave theory of quantum mechanics.

In the last frame we saw that λ_N is approximately equal to $\dfrac{h}{\sqrt{\dfrac{mZe^2}{R_N}}}$. In frame 13-5 we saw that $\lambda_N = \dfrac{4R}{N}$ for the Nth order standing wave. Now let us equate these two expressions for λ_N and solve for R_N. The result is

(1) $(R_N)_{av} = \underline{}$. 63

From frame 13-6 we saw that $KE = \dfrac{1}{2} Ze^2/R$ and $U = -Ze^2/R$; hence

$W_N = KE + U$ is $W_N = -\underline{} Ze^2/R_N$. 84

If we substitute the right hand side of Eq. 1 into this we obtain the approximate result

that $W_N = -\underline{} \dfrac{Z^2 e^4 m}{h^2 N^2}$. 99

The exact answer could be obtained using [Schroedinger's ; Planck's ; DeBroglie's] equation. The exact answer agrees with ours within a factor of 2.5. 51

Go to 13-11.

13-10
(from 13-6)

In this frame we shall go through an approximate "derivation" of the radius and energy of the hydrogen atom using the modern wave theory of the electron. Instead of plotting the potential energy of a box, we plot the potential energy between an electron and a nucleus of charge Ze. We will calculate an average wavelength λ_{av} and make the

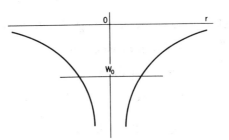

approximation that $\frac{1}{2}\lambda_{av} =$ _____ times the diameter of the hydrogen atom. 2

Or (1) $\lambda_{av} =$ _____ R_{av} where R_{av} is an average radius of the atom. 15

The average [wavelength ; half-wavelength ; radius ; diameter] should be 49

equal to $\dfrac{h}{\sqrt{2m(KE)_{av}}}$.

Hence (2) $\lambda_{av} = \dfrac{h}{\sqrt{2m(KE)_{av}}}$ [true ; false] . 20

The average KE will be _____ times the magnitude of the average potential energy: 23

$KE_{av} =$ _____ $\dfrac{Ze^2}{R_{av}}$. 30

Substitute this into the [right ; left] hand side of Eq. [1 ; 2] . 74
 23

The result is $\lambda_{av} =$ _____ . 101

Now equate this to the right hand side of Eq. [1 ; 2] and solve for R to get the 45

result $R_{av} =$ _____ . 88

This completes our derivation (with approximations) of the radius of the hydrogen atom using the modern wave theory of matter (also called quantum mechanics). Note that the Bohr theory gives the result $R = \dfrac{h^2}{4\pi^2 mZe^2}$ which is quite similar to our result.

Go to 13-9.

13-11
(from 13-9)

We shall now go through the older derivation of the radius of the hydrogen atom using the Bohr model which is about [5 ; 15 ; 30] years older than the wave theory used in the preceding frames. **65**

The Bohr postulate states that for a circular orbit the [angular momentum ; total energy ; KE] which is $R_N m v$ can only have the values $\frac{Nh}{2\pi}$; i.e., $R_N m v = \frac{Nh}{2\pi}$. **38**

(1) Solving for v gives $v = \frac{Nh}{2\pi m R_N}$

The electrostatic force is _____ . **54**

By equating [angular momentum ; KE ; magnetic force ; total energy ; centripetal force] to electrostatic force we obtain **91**

(2) $\frac{mv^2}{R_N} = \frac{Ze^2}{R_N^2}$

Substitute the right hand side of Eq. 1 into Eq. 2 and solve for R_N. The result is

$R_N =$ _____ times $\frac{h^2}{4\pi^2 Z m e^2}$. **13**

Go to 13-12.

13-12
(from 13-11)

We can obtain the energy levels of the Bohr theory by using the equation $W = KE + U$ where $KE =$ _____ $\frac{Ze^2}{R}$ and $U = -\frac{Ze^2}{R}$. Hence $W_N = -\frac{Ze^2}{2R_N}$. **21**

Now substitute the final result of the previous frame into this. The result is

$W_N = -\frac{Ze^2}{2}$ times $\left(\frac{N^2 h^2}{4\pi^2 Z m e^2}\right)$ to the _____ power. **78**

This is $W_N =$ _____ . **105**

This calculation of the hydrogen energy levels in terms of e, h, and m was first made by _____ . **94**

Go to 13-13.

13-13
(from 13-12)
Some of the possible electron clouds about a proton can have circular motion around an axis passing through the proton. This circular motion or angular momentum must be such that the number of wavelengths around the circle be one whole integer. We see that the condition of standing waves quantizes the angular momentum as well as the energy. The result is that the angular momentum can only be $\frac{\ell h}{2\pi}$, where ℓ can be any integer from 0 to $(N-1)$.

In these three-dimensional standing waves there is a third and final quantum number, m_ℓ, which can have integral values ranging from _____ to $+\ell$. Hence, for a given value of ℓ there will be _____ different possible values of m_ℓ.
m_ℓ can be zero [*true* ; *false*] . m_ℓ can have the value $(N-1)$ [*true* ; *false*] .
m_ℓ can have the value N [*true* ; *false*] .

104
64
90
42
61

Go to 13-8

13-14
(from 13-8)
For each specific combination of N, ℓ, and m_ℓ there can be a maximum of _____ electrons.

For each combination of N and ℓ there can be a total of _____ electrons (counting all possible states of m_ℓ). For the $N = 3$, $\ell = 2$ combination there can be a maximum of [2 ; 4 ; 5 ; 8 ; 6 ; 10 ; 12] electrons.

Counting all possibilities of ℓ and m_ℓ there can be a maximum of [2 ; 4 ; 6 ; 8 ; 10 ; 12] electrons in the $N = 2$ states.

28

50

59

87

If incorrect go to 13-4 ; if correct go to 13-15.

13-15 Bohr later modified his theory to account for the quantum number ℓ. He pictured orbits
(from 13-4 or 13-14) of different ℓ or angular momentum as ellipses of different eccentricities. A long, slender ellipse would have [more ; less] angular momentum than a circle of the same diameter. Orbits with the maximum ℓ - values (ℓ = $N - 1$) were represented as full circles. Pictured below are the Bohr ellipses up to $N = 3$. On each ellipse there is a box in which you should indicate the maximum number of electrons permitted in that orbit. Be sure to mark each box.

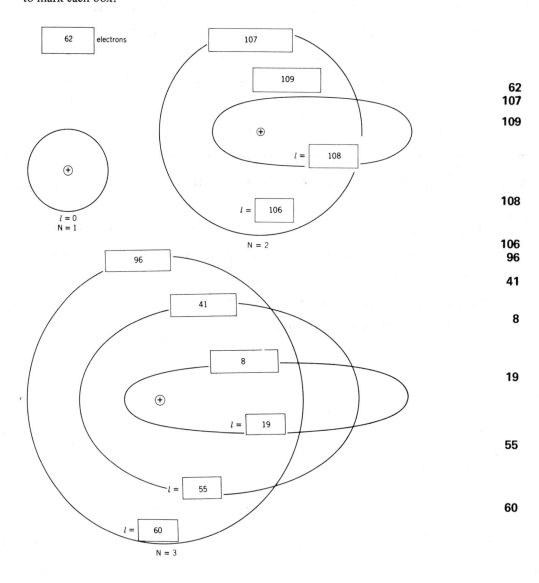

89

62
107
109

108

106
96

41

8

19

55

60

Go to 13-16.

13-16 Consider two orbits of different ℓ, but the same N. The orbit or electron cloud of the
(from 13-15) [smaller ; larger] ℓ-value gets closer to the nucleus. For the same N, electrons of [higher ; lower] ℓ will be more strongly bound. Electrons tend to "drop down" into those orbitals that are [more ; less] strongly bound.

43
66
3

Go to 13-17.

13-17 The binding energy of an outer electron is the energy required to [(a) remove the electron from the atom ; (b) bring the electron from infinity to the atom] .
(from 13-16)

In the Bohr model of the hydrogen atom the KE of the electron is _____ times 13.6 ev.

The potential energy is $\left[\dfrac{1}{2}\ ;\ 1\ ;\ 1\dfrac{1}{2}\ ;\ 2\right]$ times [13.6 ; -13.6] ev.

The total energy (KE + U) is _____ ev.

The binding energy is _____ ev.

If the electron in the hydrogen atom was at rest, but still at the distance $R = \dfrac{h^2}{4\pi^2 m e^2}$

the energy required to remove the electron would be _____ times [13.6 ; -13.6] ev and the binding energy would then be ____ times [13.6 ; -13.6] ev.

58

80

39
16

71

92

72
48
79
33

Go to 13-18.

13-18 In the periodic table we note that there is an inversion or reversal of order between the
(from 13-17) $N = 3$, $\ell = 2$ level and the $N =$ _____ , $\ell =$ _____ level. This occurred at $Z =$ _____ .

10
52
86

Go to 13-23.

203

13-19
(from 13-21)

Depending on the ℓ-values, a $N = 4$ state can be more strongly bound than a $N = 3$ state [*true* ; *false*].   98

An energy level diagram for potassium would look like Figure [A ; B].   82

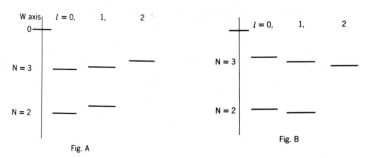

The $N = 4$, $ℓ = 0$ level would be at position [*a* ; *b* ; *c*].   25

The $N = 4$, $ℓ = 0$ level should be [*lower* ; *higher*] than the $N = 3$, $ℓ = 2$ level.   32

Go to 13-24.

13-20
(from 13-22)

Except for $ℓ = 0$, all electron clouds have zero intensity at their centers [*true* ; *false*].   5

The [3,1 ; 3,2 ; or 4,0] combination of N and $ℓ$ is most dense at the position of the nucleus.   37

Go to 13-21.

13-21
(from 13-20)

In this frame we will study in more detail the reason for the level inversion at $Z = 19$ (potassium). Potassium has _____ outer electron(s). Suppose this electron when in the $N = 3, \ell = 2$ state never penetrates any inner electron clouds. Then Z_{eff} for this outer electron would be _____ . The binding energy of this electron would be 13.6 ev times $\dfrac{Z_{\text{eff}}^2}{N^2}$ which would have the value _____ .

Suppose the outer electron of potassium when in the $N = 4, \ell = 0$ state has 10% of its cloud very near the nucleus where it "sees" a Z of 19. Assume the remaining 90% of the time it is outside all the other electron clouds. Then the average Z_{eff} for this $N = 4$ outer electron would be [1 ; 2.8 ; 9.5 ; 19] and it would have $\dfrac{Z_{\text{eff}}^2}{N^2}$ [greater ; less ; the same] than would be if this electron were in the $N = 3, \ell = 2$ state.

46

68

24

14

57

Go to 13-19.

13-22
(from 13-23)

An $\ell = 0$ electron cloud is always most dense at the position of the nucleus [*true* ; *false*] . An $\ell = 1$ cloud is always most dense at the position of the nucleus [*true* ; *false*] .

95
27

Go to 13-20.

13-23
(from 13-18)

The outer electron of lithium has a binding energy of $W = -13.6 \dfrac{Z_{\text{eff}}^2}{N^2}$ which is measured to be 4.5 ev. In this formula $N = $ _____ .

We can solve for Z_{eff} to find that it has the value _____ .

If the outer electron cloud in lithium never penetrated the inner cloud at all, then Z_{eff} would be _____ .

35

18

83

Go to 13-22.

13-24 Review Questions:

(from 13-19)

1. How many $\ell = 0$ electrons can there be in argon? _____ 7

2. How many $\ell = 1$ electrons can there be in argon? _____ 76

3. The total number of electrons that can fill up the Nth energy shell is $2N^2$, true or false? (Try this formula out for $N = 1, 2, 3,$ and 4.) _____ 34

4. How many electrons can be fitted into $N = 4, \ell = 3$? (These atoms are known as the rare earths.) _____ 103

5. In the Bohr model what is the ratio of the kinetic to potential energy?

$$\frac{KE}{U} = \underline{\quad\quad}$$ 97

Chapter 13 Answer Sheet

Item Number	Answer
1	$2R$
2	1
3	more
4	$\dfrac{4R}{N}$
5	true
6	(c)
7	6
8	2
9	2
10	4
11	6
12	true
13	N^2
14	2.8
15	4
16	-13.6
17	$4R$
18	1.25
19	zero
20	true
21	$\dfrac{1}{2}$
22	position
23	2
24	$\dfrac{1}{9}$
25	b
26	8
27	false
28	2
29	$\dfrac{h}{\sqrt{2mKE}}$
30	$\dfrac{1}{2}$
31	false
32	lower
33	13.6
34	true
35	2
36	$\dfrac{1}{2}N$
37	4, 0
38	angular momentum
39	2
40	$\dfrac{hf}{c}$
41	6
42	true
43	smaller
44	$-\dfrac{1}{2}$

Item Number	Answer
45	1
46	1
47	$2R$
48	13.6
49	wavelength
50	$2(2l+1)$
51	Schroedinger's
52	zero
53	$\dfrac{h}{p}$
54	$\dfrac{Ze^2}{R^2}$
55	1
56	true
57	greater
58	(a)
59	10
60	2
61	false
62	2
63	$\dfrac{N^2h^2}{16mZe^2}$
64	$2l+1$
65	15
66	lower
67	2
68	1
69	b
70	2
71	-13.6
72	2
73	$\dfrac{h}{\lambda}$
74	right
75	$(a+b)$
76	12
77	false
78	-1
79	2
80	1
81	0.5
82	A
83	1
84	$\dfrac{1}{2}$
85	N
86	19
87	8
88	$\dfrac{h^2}{16mZe^2}$
89	less

Item Number	Answer
90	true
91	centripetal force
92	13.6
93	$\dfrac{h}{\lambda}$
94	Bohr
95	true
96	10
97	$-\dfrac{1}{2}$
98	true
99	8
100	$\dfrac{2L}{N}$
101	$\dfrac{h}{\sqrt{\dfrac{mZe^2}{R_{av}}}}$
102	λ
103	14
104	$-l$
105	$-\dfrac{2\pi^2 mZ^2 e^4}{N^2 h^2}$
106	1
107	6
108	0
109	2

Chapter 14
THE STRUCTURE OF MATTER

14-1   A proton and electron when brought together will [release ; use up] 13.6 ev of energy.   **10**

Go to 14-2.

14-2   Two protons and two electrons can be bound together in a single hydrogen molecule.
(from 14-1)   If all four of these particles are separated by large distances from each other, they will release [4.48 ; 13.6 ; 2 × 13.6 ; (2 × 13.6 − 4.48) ; (2 × 13.6 + 4.48)] ev when brought together to form a hydrogen molecule.   **36**

Go to 14-3.

14-3   A moving hydrogen atom having kinetic energy could convert some of the kinetic
(from 14-2)   energy to ionization energy in a collision.

Suppose a hydrogen atom having KE = 15 ev collides with a stationary atom, and during the collision the hydrogen atom becomes ionized, then the remaining kinetic energy of the ionized hydrogen atom plus the energy given to the atom with which it collided would be [(15 − 13.6) ; 13.6 ; 15] ev.   **37**

Go to 14-4

14-4
(from 14-3)

In hydrogen gas (H_2) the kinetic energy in a collision could be used to convert molecular hydrogen to atomic hydrogen. In that case the initial kinetic energy would have to be greater than [4.48 ; 13.6 ; 13.6 + 4.48] ev. Hydrogen molecules will have average kinetic energies of about this amount at temperatures of about [300 ; 3000 ; 30,000] °K.

68

5

Go to 14-15.

14-5
(from 14-18)

Two moles of atomic hydrogen will release about [3,200 ; 4.3×10^{12} ; $4.48 N_0$] ergs of heat when combined to form _____ mole(s) of molecular hydrogen.

40

48

If incorrect go to 14-14 ; if correct go to 14-6.

14-6
(from 14-5 or 14-14)

The outer electrons in all covalent crystals will be "free" conduction electrons [*true* ; *false*]. All covalent crystals are insulators [*true* ; *false*]. Metallic binding is [covalent ; ionic] .

19

78

21

Go to 14-7.

14-7
(from 14-6)

The term Fermi energy usually means the [potential ; potential minus kinetic ; the average kinetic ; the maximum kinetic] energy of the conduction electrons inside a metal.

57

Go to 14-9.

14-8
(from 14-17 or 14-22)

In leaving the surface of a metal the kinetic energy of an electron is decreased by an amount equal to the work function [*true* ; *false*].

The additional kinetic energy necessary for an electron to leave the metal is the work function [*true* ; *false*].

49

77

Go to 14-12.

14-9
(from 14-7)
In this diagram [a ; b ; $a+b$] is the Fermi energy.

The work function is [a ; b ; $a+b$]. Assume an electron at the top energy level picks up an additional energy of $4kT$ in a thermal collision. The energy required to remove this electron from the metal is now _____ . This is [greater ; less] than the work function. The work function is the minimum energy required to remove an electron only when the metal is at absolute zero [true ; false]. At absolute zero all possible energy levels below the Fermi level will be occupied [true ; false].

17
39
54
26
71
46

Go to 14-17.

14-10
(from 14-16 or 14-23)
When a completely pure semiconductor is cooled to absolute zero its electrical resistance [increases to a finite limit ; decreases to a limit ; increases to infinity ; decreases to zero].

80

Go to 14-24.

14-11
(from 14-20)
The contact potential between two different metals is equal to the difference between their [Fermi energies ; work functions ; potential energies].

2

Go to 14-25.

14-12
(from 14-8)
An electron outside a metal surface has KE = 0. Its total mechanical energy (KE + U) is zero [true ; false]. The positive atomic nuclei of the metal will attract this electron [true ; false]. Suppose this attractive force gives the electron 5 ev of kinetic energy by the time it has entered the metal. The sum (KE + U) inside the metal will be _____ ev.

50
4
43

If incorrect go to 14-21 ; if correct go to 14-13.

14-13
(from 14-2 or 14-21)

Inside this metal $U = -5$ ev. If the Fermi energy of this metal is 3 ev, it will take at least _____ ev of additional energy to remove a conduction electron from the surface. Hence the work function would be _____ ev. We see that there is a simple relation between work function W, potential energy, U, and the maximum kinetic energy KE_0 of the conduction electrons. This relation is $-U =$ _____ where U is a [positive ; negative] number. In the photoelectric effect the maximum kinetic energy of the emitted electrons is related to the work function by the equation $W =$ [$KE_{max} - hf$; $hf - KE_{max}$; KE_{max}].

42
11
23
56
64

Go to 14-16.

14-14
(from 14-5)

Each molecule of H_2 formed will release $\left[4.48 \times 1.6 \times 10^{-12} \; ; \; \dfrac{1.6 \times 10^{-12}}{4.48} \; ; \; 4.48 \right]$ ergs of energy. The formation of one mole of H_2 will release $\left[\dfrac{1}{2} \; ; \; 1 \; ; \; 2 \right] \times 6.02 \times 10^{+23}$ times as much energy. To convert to calories one would [multiply ; divide] by 4.18×10^7.

73
14
30

Go to 14-6.

14-15
(from 14-4)

Hydrogen molecules at a temperature T will have an average kinetic energy per molecule equal to [$\sqrt{kT}$; $N_0 kT$; neither].

53

Go to 14-18.

14-16
(from 14-13)

If a metallic crystal is compressed by 1% in all three directions, its volume will be reduced by _____ %. The wavelength of each conduction electron in the crystal will then be [reduced ; increased] by _____%. The momentum of each conduction electron will be [increased ; decreased] by _____ %. The Fermi energy will be correspondingly [increased ; decreased] by _____ %.

9
69
16
35
47
76
55

If incorrect go to 14-23 ; if correct go to 14-10.

212

14-17
(from 14-9)
According to the theory of metals conduction electrons have kinetic energy while in the metal. For a certain metal the Fermi energy is 3 ev, but these electrons would need an additional 2 ev to leave the surface. The work function of this metal is _____ ev.   **13**
Electrons having a total kinetic energy of _____ ev would just barely be able to leave   **1**
this metal and would lose _____ ev of kinetic energy in the process of leaving the   **52**
metal. Suppose one of these 3 ev electrons absorbs a 4 ev photon. The kinetic energy of this electron will be _____ ev before leaving the surface. After leaving the surface its   **61**
kinetic energy will be _____ ev less or KE $_{final}$ = _____ ev.   **67**
  44

If incorrect go to 14-22 ; if correct go to 14-8.

14-18
(from 14-15)
In H_2 gas the average kinetic energy per [molecule ; atom] will be _____ kT. At   **7**
room temperature (T = 300°K), kT = 2.5 × [10^{-12} ; 10^{-2} ; 10^2] ev.   **51**
  32

Go to 14-5.

14-19
(from 14-25)
The wave function will curve [away from ; toward] the x-axis while inside the barrier.   **33**

Go to 14-27.

14-20
(from 14-24)
Before two metals are joined, the difference in heights between the Fermi levels is equal to the difference between their [Fermi energies ; work functions ; potential energies] .   **34**

Go to 14-11.

14-21
(from 14-12)
In this closed system of electron plus metal, mechanical energy will be conserved (assuming no friction) [*true* ; *false*]. Until this electron collides with other electrons   **38**
the quantity KE + U will be zero whether it is inside or outside the metal. Inside the metal KE = 5 ev and U = [−5 ; 0 ; 5] ev.   **15**

Go to 14-13.

14-22
(from 14-17)
In the previous problem a surface electron absorbs a photon of energy $hf = 4$ ev. If the work function $W = 2$ ev, the kinetic energy left over after leaving the metal will be [$W - hf$; $hf - W$; $hf + W$] which is _____ ev.

6
29

Go to 14-8.

14-23
(from 14-16)
Fermi energy is the [*kinetic* ; *potential*] energy of conduction electrons. Kinetic energy is proportional to the [*first power* ; *square* ; *cube*] of the momentum. $(1.01)^2 = 1.02$ [*true* ; *false*] . 1.02 is a _____ % increase from the value 1.00.

79
65
24
8

Go to 14-10.

14-24
(from 14-10)

Two different metals A and B are to be brought into contact. The occupied energy levels of the conduction electrons are shown in the first figure. The work function of A is _____ ev. The work function of B is _____ ev.

The second figure shows the situation when the metals first touch each other. At that time there would be a net flow of electrons from [A to B ; B to A ; neither]. Metal B would then acquire a [positive ; negative ; neither] charge. In a potential energy diagram the potential energy curve of B would move [up ; down] with respect to A.

(Remember these curves represent the potential energy of an electron.) The number of electrons moving from A to B which are necessary to change the potential of metal B by 1 volt happens to be very much less than 1% of the total number of conduction electrons. After reaching equilibrium the potential energy diagram would be as follows:

The potential difference between A and B would be _____ volt(s) with [A ; B] more negative. When two different metals are brought together into contact a potential difference or contact potential will be formed [true ; false].

Go to 14-20.

74
45
3
59
28
12
41
18

14-25
(from 14-11)

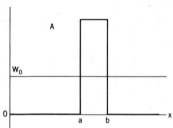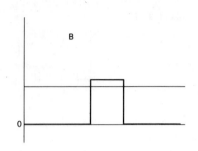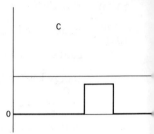

Consider a particle traveling from left to right with energy W_0. This particle will encounter the potential barrier shown above in either A, B, or C.

If the wave function of the particle is as shown to the right, it must have encountered barrier [A ; B ; C] .

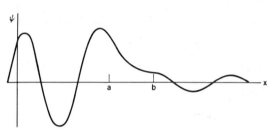

22

If the wave function is as shown to the right, the particle must have encountered barrier [A ; B ; C] .

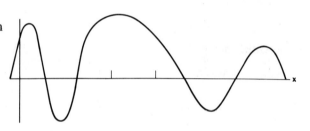

72

If the wave function is as shown to the right, the particle must have encountered barrier [A ; B ; C] .

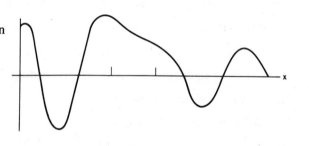

58

Go to 14-19.

14-26 An electron gains 9 ev of kinetic energy as it enters the surface of a metal. For this metal
(from 14-27) the [(a) Fermi energy ; (b) depth of the potential well ; (c) work function] is 9 ev.   66

Go to 14-28.

14-27
(from 14-19)

While inside the barrier the wave function drops off exponentially. This means if ψ decreases by a factor f in a distance Δx, it will decrease by another factor f in the second interval of Δx, and so on.

Consider a barrier 1 cm long for which the probability of barrier penetration is $\frac{1}{16}$.

The ratio of $\frac{\psi_{out}}{\psi_{in}} =$ _____ . Suppose this barrier is increased to 2 cm in length.   62

Then the probability of barrier penetration will be _____ and the ratio $\frac{\psi_{out}}{\psi_{in}}$ 20

will be _____ . If this barrier is increased to a 3 cm length, $\frac{\psi_{out}}{\psi_{in}}$ will be   31

_____ .   75

Go to 14-26.

14-28
(from 14-26)

Review Questions:

1. A particular metal has a work function of 3 ev and a maximum Fermi energy of 5 ev. What is the depth of the potential well? _____   27

2. In a metal at room temperature there will be some unoccupied energy levels below the Fermi level. True or false? _____   60

3. Write a formula for the kinetic energy of an electron in terms of its wavelength.

   KE = _____   70

4. Two different metals which happen to have identical potential wells have Fermi levels differing by 2 ev. What will be the contact potential between these two metals?

   _____   25

5. In the above problem will the metal with lowest Fermi level be positive or negative relative to the other? _____   63

Chapter 14 Answer Sheet

Item Number	Answer	Item Number	Answer
1	5	43	zero
2	work functions	44	2
3	A to B	45	3
4	true	46	true
5	30,000	47	1
6	$hf - W$	48	1
7	molecule	49	false
8	2	50	true
9	3	51	$\dfrac{3}{2}$
10	release	52	5
11	2	53	neither
12	1	54	$a - 4kT$
13	2	55	2
14	1	56	negative
15	-5	57	maximum kinetic
16	1	58	B
17	b	59	negative
18	true	60	true
19	false	61	7
20	$\dfrac{1}{256}$	62	$\dfrac{1}{4}$
21	covalent	63	negative
22	A	64	$hf - KE_{max}$
23	$KE_0 + W$	65	square
24	true	66	(b)
25	2 ev	67	5
26	less	68	4.48
27	8 ev	69	reduced
28	up	70	$\dfrac{h^2}{2m\lambda^2}$
29	2	71	true
30	divide	72	C
31	$\dfrac{1}{16}$	73	$4.48 \times 1.6 \times 10^{-12}$
32	10^{-2}	74	2
33	away from	75	$\dfrac{1}{64}$
34	work functions	76	increased
35	increased	77	true
36	$2 \times 13.6 + 4.48$	78	false
37	$15 - 13.6$	79	kinetic
38	true	80	increases to infinity
39	a		
40	4.3×10^{12}		
41	B		
42	2		

Chapter 15
NUCLEAR PHYSICS

15-1 It is a good approximation that the atomic weight of an isotope of mass number A is

$\left[A \; ; \; \frac{1}{2}A \; ; \; AN_0 \right]$. **3**

If an element consists of an equal abundance of two isotopes, one having $A = 30$ and the other $A = 28$, the atomic weight of this element will be approximately _____ . **15**

The mass of any nucleus (except hydrogen) will be [greater than ; less than ; the same as] AM_P where M_P is the mass of the proton. **32**

If M_A is the mass of a nucleus having a mass number equal to A, the fractional difference $\dfrac{AM_P - M_A}{M_A}$ is greatest for [H ; He ; C ; Fe ; Pb ; U] . **50**

Go to 15-3.

15-2
(from 15-6)
On any collision the chance of escaping is always $\dfrac{\psi_{out}}{\psi_{in}}$ [*true* ; *false*]. **41**

If incorrect go to 15-4 ; if correct go to 15-8.

221

15-3
(from 15-1)

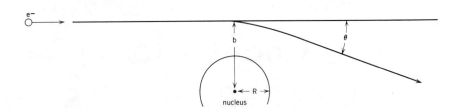

In this frame we shall show how the radius of an atomic nucleus can be determined by scattering high energy electrons against thin foils. In the figure it is seen that an electron with a certain impact parameter b will be scattered by an angle θ. Because of Coulomb's law θ is simply related to b. As b decreases, θ will [increase ; decrease]. Inside the nucleus the electric field will be [stronger ; weaker] than what it is at the surface of the nucleus. The maximum scattering angle θ_{max} corresponds to $b =$ [R ; $2R$; 1.2×10^{-13} cm]. It can be shown that the relation between b and θ for high energy electrons is $b = \dfrac{Ze^2}{W\tan\frac{1}{2}\theta}$ where W is the electron energy.

73
48
59

Let N_θ be the number of electrons scattered by an angle greater than θ. Plotted as a function of θ, this will look like curve _____ .

12

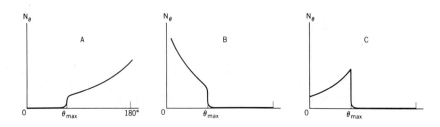

Go to 15-7.

15-4
(from 15-2)

On any collision the chance of escaping is $\dfrac{\psi^2_{out}}{\psi^2_{in}}$ [true ; false].

1

Go to 15-8.

222

15-5
(from 15-7)
Another method for determining the size of the nucleus is to shoot neutrons at atoms. If a neutron hits a nucleus of cross sectional area σ, it will interact and be lost from the beam. Let N_a be the number of atoms in a sheet of copper and let a be the area of the sheet of copper. Then the chance that any one neutron will not pass through is _____ where N is the number of neutrons hitting the sheet. The number of neutrons lost from the beam is $N - N' =$ _____. Now solve the above equations for σ. The result is $\sigma =$ _____. The number of atoms in the sheet is $\left[\dfrac{M}{A} \; ; \; \dfrac{A}{M} \; ; \; AM \right]$ times N_0 (Avogadro's number) where M is the mass of the sheet and A is the [atomic number ; mass number] . Experiments gives the result $\sigma = 0.5 \times 10^{-24}$ cm^2 for copper. The corresponding radius would be $R =$ _____ $\times 10^{-13}$ cm.

7
46
18
27
33
75

Go to 15-6.

15-6
(from 15-5)
An alpha particle is prevented from escaping from a nucleus by a potential barrier. Assume for a particular isotope an alpha particle will escape after 10^{20} collisions on the average. The chance for it to escape during the first 10^{10} collisions will be $\dfrac{1}{2}$ the chance for it to escape during 10^{20} collisions [*true* ; *false*]. The chance for an alpha particle escaping on the 10^{20}-th collision will be exactly the same as its chance of escaping on the first collision [*true* ; *false*]. The chance of escaping on any particular collision will be the same as its chance of escaping on the first collision [*true* ; *false*].
One half of 10^{20} is 10^{10} [*true* ; *false*].

82
54
23
10

Go to 15-2.

15-7
(from 15-3)
A beam of 100 Mev electrons is shot at a copper foil ($Z = 29$). Scatterings up to $16°$ are observed. From this data we can determine the radius of the copper nucleus [*true* ; *false*]. The radius is given by the formula $R = \dfrac{Ze^2}{W \tan 8°}$ [*true* ; *false*].

The electron energy W is _____ ergs.

Using the value $\tan 8° = 0.14$, the final result will be $R =$ _____ cm.

2
9
39
13

Go to 15-5.

15-8
(from 15-2 or 15-4)
In radioactive decay, if the probability of decay per unit time is doubled, the halflife will be doubled [*true* ; *false*].

82

Go to 15-9.

15-9
(from 15-8)
A radioactive isotope with a one year halflife is produced. Consider an individual nucleus that doesn't decay in the first year. The probability that this nucleus will decay sometime during the second year is _____ .

55

If this particular nucleus happens to survive the first two years without decaying, the probability that it will decay sometime during the third year is _____ .

14

Go to 15-11.

15-10
(from 15-11)
Pu^{239} has less mass than U^{239} [*true* ; *false*].

22

U^{239} has less mass than U^{238} plus a free neutron [*true* ; *false*].

62

Go to 15-19.

15-11
(from 15-9)
A radioactive sample of one year halflife is prepared. The fraction of nuclei surviving the first year will be _____ . The fraction surviving the first two years will be _____ .

71
40

Let N_0 be the initial number of radioactive nuclei. Let N' be the number remaining after the first 6 months and N'' the number remaining after the first year. The fraction decaying in each 6 month period is always the same [*true* ; *false*]. Then

57

$$\frac{N''}{N'} = \frac{N'}{N_0} \quad \text{or} \quad N'^2 = N''N_0 \quad [\text{ true ; false }].$$

76

The fraction decaying in one year is $\frac{N''}{N_0}$ which equals _____ .

16

We now have two simultaneous equations and can eliminate N''. Then we obtain the relation $\frac{1}{2}N_0^2 =$ [$(N')^2$; N' ; $\sqrt{N'}$] . The final result is $N' \approx$ _____ % of N_0.

30
44

The number surviving the first 4 months will be $\left[\dfrac{1}{\sqrt[4]{2}} ; \dfrac{1}{\sqrt[3]{2}} ; \dfrac{1}{\sqrt{2}} ; \dfrac{1}{3} ; \dfrac{1}{\sqrt{3}} ; \text{neither} \right]$ times N_0.

5

The number surviving the first 3 months will be N_0 divided by [2 ; 3 ; 4 ; $2^{\frac{1}{2}}$; $2^{\frac{1}{3}}$; $2^{\frac{1}{4}}$] .

21

Go to 15-10.

15-12
(from 15-17)

The difference in binding energy/nucleon between He⁴ and Li⁶ is 1.9 Mev/nucleon [*true* ; *false*] . The rest energy per nucleon is Mc^2 or about [100 ; 200 ; 930 ; 1830] Mev. The ratio of the above two numbers will be the fractional energy release per nucleon [*true* ; *false*] . The ratio is _____ %.

66
84

19
63

Go to 15-18.

15-13
(from 15-19)

The wavelength of the neutron bound to the proton in the deuterium nucleus is approximately given by $\left[\frac{1}{4}\lambda \; ; \frac{1}{2}\lambda \; ; \lambda\right] = R$ where R is the radius of the potential well. Hence $\lambda =$ [1 ; 2.5 ; 5 ; 10] × 10⁻¹³ cm. The effective momentum of the neutron bound inside the deuteron can be obtained from the relation $P = \frac{h}{\lambda}$ [*true* ; *false*] . The corresponding kinetic energy would be $\frac{P^2}{2M}$ [*true* ; *false*] .

The magnitude of the depth of the potential well would be this kinetic energy [*plus* ; *minus*] the measured binding energy. This is how the "strength" of the neutron-proton force was originally determined [*true* ; *false*].

4

70

38

11

81
64

Go to 15-14.

15-14
(from 15-13)

The previous potential well can contain [0 ; 1 ; 2 ; 3] standing wave(s). If the well were made deeper, it could hold more standing waves [*true* ; *false*] . If the depth of the well was kept the same, but the radius R was increased, the well could hold more standing waves [*true* ; *false*]. The potential well corresponding to a large nucleus is deeper [*true* ; *false*]. It also has a larger R [*true* ; *false*]. Then the number of possible bound states will exceed the number of neutrons (or protons) in the nucleus [*true* ; *false*].

Normally all the lowest bound states are completely filled with [neutrons ; electrons] . Consider two nuclei having the same depth of potential well, but different radii such as He and Pb. In He the spacing between the energy levels will be [greater ; less] than that of Pb. The relative spacing of the neutron energy levels is the same as for the electron energy levels in hydrogen; i.e., it goes as $\frac{1}{N^2}$ [*true* ; *false*] .

17
28

49
78
35

65

42

20

83

Go to 15-15.

15-15
(from 15-14)

Consider a one-dimensional model of a nucleus consisting of a one-dimensional potential well. (We only consider waves that travel in the *x*-direction).

If the 10th energy level ($N = 10$) is filled up with neutrons, but $N = 11$ is empty, this "nucleus" will contain _____ neutrons. If a neutron is excited from the $N = 10$ to $N = 11$ energy level, its wavelength will [increase ; decrease] by about _____ %. And its momentum will [increase ; decrease] by about _____ %. Its kinetic energy will [increase ; decrease] by about _____ %.

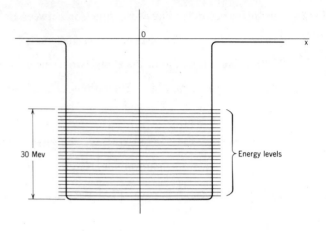

30 Mev

Energy levels

36
8
74
43
51
26
60

Suppose that the Fermi level ($N = 10$) has a kinetic energy of 30 Mev. Then the kinetic energy of a neutron in the first excited state ($N = 11$) will be _____ Mev. The spacing in Mev between the first excited state and the Fermi level will be _____ Mev.

79
53

If the radius of this "nucleus" is doubled, but the depth of the well kept the same, the Fermi energy of 30 Mev would correspond to $N =$ _____ .

29

The first excited state would then be about _____ % higher in energy than the 30 Mev Fermi level. The spacing in Mev between these two energy levels would be about _____ Mev.

68
72

Suppose the depth of the potential well is 40 Mev. Then the neutron binding energy will be _____ Mev.

58

Go to 15-16.

15-16
(from 15-15)

In fission a heavy nucleus such as uranium splits up into [2 ; 3 ; 4] smaller nuclei with an energy release of about [1 ; 10 ; 100 ; 200 ; 1000] Mev. Each fission fragment will have a Z of about [25 ; 45 ; 65 ; 80] and an A of about [20 ; 60 ; 110 ; 180] . The sum of the two radii would be about [1 ; 2 ; 10 ; 20] × 10^{-13} cm. When touching each other the electrostatic energy of repulsion would be about $\frac{Q_1 Q_2}{R}$ where $Q_1 = Q_2 =$ [2 × 10^{-8} ; 4.8 × 10^{-10} ; 40] statcoulombs. This is an energy of _____ ergs or about [100 ; 200 ; 250 ; 500] Mev. Assume these two fission products can be up to 10^{-12} cm apart in the parent neucleus, then the potential barrier for fission would be about [2 ; 10 ; 50 ; 100 ; 300] Mev.

90
87
89
86
85
88
67
80
24

Go to 15-17.

15-17
(from 15-16)

A common thermonuclear reaction is $N + Li^6 \rightarrow H^3 + X$ where X is _____ (N stands for neutron). The H^3 which is [deuterium ; tritium] then gets converted into He^4. by other reactions. The end result is a given mass of Li^6 converted into He^4. If the binding energy in Mev/neucleon is 5.2 Mev for Li^6 and 7.1 Mev/nucleon for He^4, the percent of rest mass "converted" into energy is [0.1 ; 0.2 ; 0.6] %.

6
25
52

If incorrect go to 15-12 ; if correct go to 15-18.

15-18
(from 15-17 or 15-12)

A proton in thermal equilibrium with interstellar gas clouds of 1 gm mass and average velocity of 1 m/sec would have an energy of about [0.5 ; 1 ; 50 ; 5000] ergs. This corresponds to a cosmic ray energy of about [3 × 10^2 ; 3 × 10^6 ; 3 × 10^8] Bev.

31
45

Go to 15-20.

15-19
(from 15-10)
Assume that it takes 30 ev of radiation on the average to produce one ion pair. The number of ion pairs per gram produced by one roentgen of radiation is $\left[\dfrac{83}{30} \; ; \; \dfrac{83N_0}{30} \; ; \; \dfrac{83}{30 \times 1.6 \times 10^{-12}}\right]$.

37

77
69

The amount of negative charge in statcoulombs liberated would be the previous number times _____ . The final result is _____ statcoulombs.

Go to 15-13.

15-20 Review Questions:

(from 15-18)
1. After 4 halflives what percent of a sample remains? _____ %.   34
2. Is the neutron heavier or lighter than a proton plus electron? _____   47
3. Consider a nucleus containing 30 protons. If the depth of the potential well is 36 Mev and the proton binding energy is 7 Mev, what is the proton Fermi energy?

$$KE_0 = \text{\underline{\qquad}} \text{ Mev.}$$

61

4. The binding energy per nucleon is 7.1 Mev for He^4 and 2.2 Mev for H^2. What is the percent of rest mass converted into energy for a thermonuclear process that converts deuterium to helium? _____ %.

56

Chapter 15 Answer Sheet

Item Number	Answer
1	true
2	true
3	A
4	$\frac{1}{4}\lambda$
5	$\frac{1}{\sqrt[3]{2}}$
6	He^4
7	$\frac{N_a\sigma}{a}$
8	decrease
9	true
10	false
11	true
12	B
13	3.0×10^{-13}
14	$\frac{1}{2}$
15	29
16	$\frac{1}{2}$
17	1
18	$\frac{(N-N')a}{NN_a}$
19	true
20	greater
21	$2^{\frac{1}{4}}$
22	true
23	true
24	50
25	tritium
26	increase
27	$\frac{M}{A}$
28	true
29	20
30	$(N')^2$
31	5000
32	less than
33	mass number
34	6.25
35	true
36	20
37	$\frac{83}{30 \times 1.6 \times 10^{-12}}$
38	true
39	1.6×10^{-4}
40	$\frac{1}{4}$
41	false

Item Number	Answer
42	neutrons
43	increase
44	70
45	3×10^6
46	$\frac{NN_a\sigma}{a}$
47	heavier
48	weaker
49	true
50	Fe
51	10
52	0.2
53	6
54	true
55	$\frac{1}{2}$
56	.52
57	true
58	10
59	R
60	20
61	29
62	true
63	0.2
64	true
65	true
66	true
67	4×10^{-4}
68	10
69	830
70	10
71	$\frac{1}{2}$
72	3
73	increase
74	10
75	4.0
76	true
77	4.8×10^{-10}
78	true
79	36
80	250
81	plus
82	false
83	false
84	930
85	10
86	110
87	200
88	2×10^{-8}
89	45
90	2

Chapter 16
PARTICLE PHYSICS

16-1   A grain of sand contains _____ different kinds of elementary particles.   **98**

If incorrect go to 16-16 ; if correct go to 16-2.

16-2   The attractive forces between atoms that hold atoms together in ordinary matter are due
(from   to the [nuclear ; electromagnetic ; weak ; gravitational] interaction. The   **114**
16-1   **103**
or   _____ interaction holds the protons and [electrons ; neutrons] together in the   **30**
16-16)   atomic nucleus. The _____ interaction is also called the universal Fermi interaction. **110**

The _____ interaction is also called the strong interaction.   **97**

The _____ interaction accounts for the productions of pions. The _____   **111**

_____ interaction accounts for the production of beta "rays" in beta decay.   **90**

Go to 16-3.

231

16-3
(from 16-2) they a

Only certain elementary particles are capable of interacting with each other strongly by means of the nuclear or strong interaction. Classify the following particles as to whether they are strongly interacting or weakly interacting. Make a check mark in the appropriate column.

Particle	Can Interact Strongly	Cannot Interact Strongly	
pion			36
muon			71
electron			57
lambda			69
K meson			81
neutrino			94

Go to 16-4.

16-4
(from 16-3)

Classify the following reactions as to the type of interaction taking place; make a check mark in the appropriate column.

Reaction	Strong	Electromagnetic	Weak	
$\mu^+ \rightarrow e^+ + \nu + \bar{\nu}$				83
$\pi^+ \rightarrow \mu^+ + \nu$				43
$e^- + P \rightarrow \gamma + H(\text{hydrogen})$				78
$P + \bar{P} \rightarrow \Lambda + \bar{\Lambda}$				89
$\gamma \rightarrow e^+ + e^-$				52
$\bar{N} + P \rightarrow \pi^+ + \pi^0$				60
$U^{238} + N \rightarrow U^{239}$				39
$U^{239} \rightarrow Np^{239} + e^- + \bar{\nu}$				68

Go to 16-5.

16-5
(from 16-4)
In the reaction $P + P \rightarrow \overline{N} + N + P + P$ the heavy particle number on the right hand side is _____ . The heavy particle number on the left hand side is the [same ; different] .

73
13

In the reaction $P + P \rightarrow \overline{\Lambda} + \Sigma^+ + \pi^+$ the heavy particle number on the right hand side is _____ and the heavy particle number on the left side is _____ . This reaction would be [permitted ; forbidden] . It would violate the law of conservation of heavy particles [*true* ; *false*] .

49
85
22
50

In the reaction $P + P \rightarrow \overline{\Lambda} + \Sigma^+ + P + N$ the heavy particle number on the right side would be _____ . This reaction would be [permitted ; forbidden] .

31
12

Go to 16-7.

16-6
(from 16-12)
Suppose a certain radioisotope is observed always to decay emitting a beta ray along its spin direction. In the mirror image of this the beta ray would appear to be emitted [along; against] the spin direction of the nucleus. Then the mirror image result would contradict the experimental observation [*true* ; *false*] . Nuclei which emitted beta rays preferentially along their spin direction would violate conservation of parity [*true* ; *false*] .

99

46

86

Go to 16-13.

16-7
(from 16-5)

The e^- has lepton number _____ .

The e^+ has lepton number _____ .

The μ^- has lepton number _____ . The μ^+ has lepton number _____ . The decay $\mu^+ \rightarrow e^+ + \nu + \bar{\nu}$ has lepton number _____ on the left hand side. The total lepton number on the right hand side is _____ . According to the law of conservation of leptons, this reaction is [permitted ; forbidden] .

The reaction $\bar{\nu} + P \rightarrow N + e^+$ has lepton number _____ on [both sides ; left side ; right side] . This reaction is [permitted ; forbidden] . The reaction $\nu + P \rightarrow N + e^+$ has lepton number _____ on the left hand side. This reaction is [permitted ; forbidden] .

In the reaction $\mu^- + P \rightarrow N + (\nu$ or $\bar{\nu})$, the lepton number on the left side is _____ . The correct lepton on the right side will be [ν ; $\bar{\nu}$] .

2
18
45
14
26
74
10
58
61
95
84
40
79
48

Go to 16-9.

16-8
(from 16-9)

The Ξ^- hyperon decays as follows: $\Xi^- \rightarrow \pi^- + \Lambda$

An anti-Ξ^- is [positive ; negative ; neutral] in charge.

It decays into a [positive ; negative ; neutral] pion and a [lambda ; anti-lambda] . Ultimately the decay products of an anti-Ξ^- must result in an antiproton [*true* ; *false*] .

47
15
20
11

Go to 16-11.

16-9
(from 16-7)

The following decay modes are forbidden because they violate one of the conservation laws. For each decay indicate which conservation law is violated by checking the correct column.

Decays	Charge	Heavy Particles	Energy	Leptons	
$P \rightarrow e^+ + \nu$					16
$N \rightarrow e^+ + P + \nu$					4
$\Lambda \rightarrow \pi^+ + \pi^-$					23
$K^+ \rightarrow \pi^+ + \pi^0 + \pi^-$					37
$P \rightarrow N + e^+ + \nu$					9
$N \rightarrow e^- + P + \nu$					76
$H^3 \rightarrow He^3 + e^- + \nu$					41

Go to 16-8.

16-10
(from 16-13)

Just as particles have heavy particle numbers and lepton numbers, so do they have strangeness numbers. In the following reactions all particles have zero strangeness except the K^+ which has strangeness number $= +1$, the Λ and K^- which have strangeness numbers $= -1$, and the Ξ^- which has strangeness number $= -2$. In the following reactions, determine whether strangeness number is conserved.

Reaction	Strangeness Conserved Yes	No
$\pi^+ + N \to \Lambda + K^+$			106
$\pi^+ + P \to K^+ + P$			72
$\pi^- + P \to N + K^+ + K^-$			112
$K^- + P \to K^+ + \Xi^-$			87
$P + P \to \Lambda + P + K^+$			116
$\Xi^- \to \pi^- + \Lambda$			62

The strong interaction must obey conservation of strangeness [*true ; false*] . The weak interaction must obey conservation of strangeness [*true ; false*] .

66
93

Go to 16-17.

16-11
(from 16-8)

Classify the following particles by making a check in the appropriate column.

	Lepton	Meson	Hyperon
μ				54
π				59
K^0				25
ν				44
e				6
Σ^0				29
Ξ^0				63
Λ				34

Go to 16-12.

16-12
(from 16-11)
We shall use a right hand rule for angular momentum to indicate spin direction along the axis of rotation. For example, the spin direction of the earth would be along the earth's axis pointing toward the [north ; south] pole.

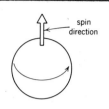

35

Suppose the spin direction of an electron is pointing along its direction of motion. Now place a mirror below the electron as shown. The mirror image of the electron will be traveling in the [same ; opposite] direction. 100

The mirror image will be rotating in the [same ; opposite] sense. The mirror image will have its spin direction [parallel ; antiparallel ; neither] to the direction of motion.

77
80

Go to 16-6.

16-13
(from 16-6)
Looking from above, the nucleus is rotating [clockwise ; counterclockwise] and its spin direction is [up ; down]. The mirror image of the nucleus is rotating [clockwise ; counterclockwise] and its spin direction is [up ; down]. In the mirror image, the beta ray would be pointing opposite to the spin direction rather than along it and this would contradict what is observed.

88

109

115
102

If the mirror image result of an experiment looks different from the experiment, then parity is not conserved [true ; false].

108

Go to 16-10.

16-14
(from 16-17)

Which of the following are true and which false?

The free neutron is unstable. _____ .   42

The free antiproton is unstable and has a halflife. _____ .   7

The free antineutron is unstable. _____ .   91

According to the principle of antiparticle symmetry the antineutron should have the same halflife as the neutron. _____ .   51

Go to 16-15.

16-15
(from 16-1)

In June 1962 experiments on detection of high energy neutrinos at Brookhaven National Laboratory determined that there are two different kinds of neutrinos. One kind, the v_e and $\bar{v}_e$, is always associated with electrons, and the other kind, the v_μ and $\bar{v}_\mu$, is always associated with muons. Hence the law of conservation of leptons now becomes two separate conservation laws: electron-type lepton conservation, and muon-type lepton conservation.

In the reaction $v_e + P \rightarrow N + e^+$, the electron-type lepton number of the left side is _____ and the electron-type number on the right side is _____ . This reaction is [permitted ; forbidden] .   1 / 5 / 17

In the following reactions choose the correct neutrino:

$v + P \rightarrow N + e^+$ $v = [\ v_e\ ;\ \bar{v}_e\ ;\ v_\mu\ ;\ \bar{v}_\mu\ ]$ 55

$v + N \rightarrow P + \mu^-$ $v = [\ v_e\ ;\ \bar{v}_e\ ;\ v_\mu\ ;\ \bar{v}_\mu\ ]$ 82

$v + N \rightarrow P + e^-$ $v = [\ v_e\ ;\ \bar{v}_e\ ;\ v_\mu\ ;\ \bar{v}_\mu\ ]$ 27

In μ^+ decay; $\mu^+ \rightarrow e^+ + v + \bar{v}$.

The neutrino is a [v_e ; v_μ] and the antineutrino is a [$\bar{v}_e$; $\bar{v}_\mu$] .   65 / 113

Go to 16-18.

16-16
(from 16-1)

A grain of sand contains protons and neutrons _____ . A grain of sand contains mesons _____ . A grain of sand contains hyperons _____ . A grain of sand contains electrons _____ .   101 / 105 / 56 / 24

Go to 16-2.

16-17	The neutrino always has its spin direction [parallel ; antiparallel] to its direction of motion. It is [left ; right] handed. The mirror image of a neutrino is always [left ; right] handed. This mirror image contradicts what is observed [*true* ; *false*] . The observation of a neutrino all by itself violates conservation of parity [*true* ; *false*] . If we charge conjugate a neutrino we obtain a [left ; right] handed [neutrino ; antineutrino] . This contradicts what is observed [*true* ; *false*] . The neutrino all by itself violates the principle of antiparticle symmetry [*true* ; *false*] . If we make both of the above operations on a neutrino; i.e., we charge conjugate the mirror image of a neutrino, we obtain a [left ; right] handed [neutrino ; antineutrino] . This contradicts what is observed [*true* ; *false*] .	104 107 19 33 75 64 53 21 3 28 92 70
(*from 16-10*)		

Go to 16-14.

16-18 Review Questions

(*from 16-15*)

1. Which of the following particles are stable?

   ($\gamma, \nu, e, \mu, \pi, K, P, N, \Lambda, \Sigma, \Xi, \Omega$). _____ 8

2. Which of these particles can interact by the strong interaction? _____ 32

3. Which are leptons? _____ 67

4. Which are strange particles; i.e., have nonzero strangeness? _____ 38

5. Which are heavy particles? _____ 96

Chapter 16 Answer Sheet

Item Number	Answer	Item Number	Answer	Item Number	Answer
1	1	56	false	110	weak
2	1	57	cannot	111	nuclear
3	true	58	-1	112	yes
4	charge	59	meson	113	$\bar{v}_\mu$
5	-1	60	strong	114	electromagnetic
6	lepton	61	both sides	115	counterclockwise
7	false	62	no	116	yes
8	γ, v, e, P	63	hyperon		
9	energy	64	left		
10	permitted	65	v_e		
11	true	66	true		
12	permitted	67	v, e, μ		
13	same	68	weak		
14	-1	69	can		
15	positive	70	false		
16	heavy particles	71	cannot		
17	forbidden	72	no		
18	-1	73	2		
19	right	74	-1		
20	anti-lambda	75	true		
21	true	76	leptons		
22	forbidden	77	opposite		
23	heavy particles	78	electromagnetic		
24	true	79	1		
25	meson	80	antiparallel		
26	-1	81	can		
27	v_e	82	v_μ		
28	right	83	weak		
29	hyperon	84	1		
30	neutrons	85	2		
31	2	86	true		
32	$\pi, K, P, N, \Lambda, \Sigma, \Xi, \Omega$	87	yes		
33	true	88	counterclockwise		
34	hyperon	89	strong		
35	north	90	weak		
36	can	91	true		
37	charge	92	antineutrino		
38	$K, \Lambda, \Sigma, \Xi, \Omega$	93	false		
39	strong	94	cannot		
40	forbidden	95	permitted		
41	leptons	96	$P, N, \Lambda, \Sigma, \Xi, \Omega$		
42	true	97	nuclear		
43	weak	98	3		
44	lepton	99	against		
45	1	100	same		
46	true	101	true		
47	positive	102	up		
48	v	103	nuclear		
49	0	104	antiparallel		
50	true	105	false		
51	true	106	yes		
52	electromagnetic	107	left		
53	antineutrino	108	true		
54	lepton	109	up		
55	$\bar{v}_e$				

Chapter 1 Answer Sheet

Item Number	Answer
1	10^3
2	$\dfrac{10^{-8}}{10^{10}}$
3	$10^{-10} \times \dfrac{1}{5}$
4	2×10^{-6}
5	6, 9, 10
6	10^{-18}
7	9^2
8	4
9	4
10	$\dfrac{AB}{A+B}$
11	$X^{1.5}$
12	$\dfrac{1}{10^4}$
13	left
14	0.000025
15	$\dfrac{1}{A+B}$
16	none
17	10^{-18}
18	2
19	$.25 \times 10^{-4}$
20	none
21	$\dfrac{1}{3}$
22	left
23	cube root of A
24	10^{-18}
25	9^2
26	.25
27	4
28	5
29	$5\sqrt{3}$
30	2.5×10^{-5}
31	true
32	10^{-9}
33	$\dfrac{B-A}{AB}$

Chapter 2 Answer Sheet

Item Number	Answer
1	d
2	false
3	a
4	45
5	15
6	35
7	$\frac{5}{6}$ mi
8	b
9	a
10	$\frac{7}{6}$
11	$s = \bar{v}t$
12	$\frac{7}{6}$ mi
13	$\frac{88}{120}$ or 0.73
14	$\frac{1}{3}$ mi
15	zero
16	$\frac{2}{60}$
17	$\frac{20 + 50}{2}$
18	9.8 m/sec²
19	false
20	20
21	$6\sqrt{3}$
22	50
23	88 ft/sec
24	$\frac{1}{3} \times \frac{88}{60}$
25	$\frac{88}{120}$
26	120 sec
27	N
28	$\frac{88}{120}$
29	false
30	$\frac{2}{3} \times \frac{88}{60}$
31	b
32	$\frac{2}{9}$
33	$\frac{1}{2} \times \frac{88}{60}$
34	a, b, and c
35	−44
36	1000

Item Number	Answer
37	a
38	625
39	$V_0 T \cos\theta$
40	shorter
41	$1000\sqrt{2}$
42	$\frac{1}{2} T$
43	21
44	2
45	$v^2 = 2as$
46	$V_x T$
47	18
48	$V_0 \cos\theta$
49	$V_0 \sin\theta = g\left(\frac{1}{2} T\right)$
50	$\frac{4}{18}$
51	$\frac{2V_0^2}{g} \sin\theta \cos\theta$
52	90°
53	constant
54	1
55	6
56	toward center of earth
57	45°
58	$\frac{1}{4} g$
59	$\frac{1}{\sqrt{2}}$
60	$\frac{V_0^2}{g}$
61	$\frac{1}{4}$
62	false
63	$z \cos\theta$
64	$\frac{\frac{1}{2} v^2}{R_e}$
65	$2\sqrt{2}$
66	$\frac{v^2}{r}$
67	$\sqrt{8}$
68	$2\pi R$
69	$\frac{1}{2}$
70	increase
71	cm
72	$\frac{2\pi R}{v}$

Item Number	Answer
73	$\frac{1}{T}$
74	$v_c \sqrt{\frac{R_e}{R_e + h}}$
75	false
76	$\frac{4\pi^2}{(465 T_0)^2} (60 R_e)$
77	$(2\pi f)^2 r$
78	$4\pi^2 f^2 r$
79	true
80	true
81	true
82	$\frac{2\pi R}{V}$
83	$\frac{(R_e + h_1)^3}{(R_e + h_2)^3}$
84	$2\pi \sqrt{\frac{(R_e + h)^3}{g R_e^2}}$
85	$\frac{4\pi^2}{T^2} (R_e + h)$
86	$2\pi (R_e + h)$

Chapter 3 Answer Sheet

Item Number	Answer
1	true
2	zero
3	$Mg \tan \theta$
4	net force
5	t_2
6	$\dfrac{2\pi x_0}{T}$
7	continue on at 60 mph
8	true
9	Ma
10	$M_2 g$
11	7.74×10^{-15}
12	zero
13	true
14	F_{app}
15	B
16	true
17	$2\pi \sqrt{\dfrac{L \cos\theta}{g}}$
18	gm/sec^2
19	$M_2 - M_2'$
20	false
21	true
22	false
23	$-\dfrac{kx}{M}$
24	true
25	$F_2 \sin\theta$
26	zero
27	1.1×10^{-3}
28	true
29	$\dfrac{M_2 g}{M_1 + M_2}$
30	true
31	true
32	Third
33	tension
34	$\tan\theta$
35	gm cm/sec^2
36	the same as
37	200
38	$\dfrac{a}{g}$
39	$\dfrac{2\pi}{T}\sqrt{x_0^2 - x^2}$
40	$M_2 g$

Item Number	Answer
41	true
42	$Mg \sin\theta$
43	$\tan\theta$
44	false
45	true
46	$\dfrac{M_1}{M_2}$
47	$\dfrac{k}{m}$
48	true
49	$\vec{F}_3$
50	Mg
51	less than
52	true
53	B
54	4
55	$\dfrac{y}{x_0}$
56	false
57	true
58	$g \sin\theta$
59	true
60	V_x
61	true
62	zero
63	C
64	$.66 \times 10^{30}$
65	1.29×10^{14}
66	repulsive
67	sec^2
68	B
69	toward the center
70	400
71	true
72	true
73	true
74	F_1
75	$-\vec{F}_3$
76	none
77	true
78	zero
79	true
80	up, but perpendicular to the table
81	false

Chapter 4 Answer Sheet

Item Number	Answer
1	4
2	false
3	$\dfrac{1}{4}$
4	true
5	true
6	$\dfrac{GM_j}{R_2}$
7	induction
8	$\sqrt{\dfrac{R_1}{R_2}}$
9	8
10	1
11	1.9×10^{27}
12	true
13	$4g$
14	false
15	$\sqrt{\dfrac{GM_e}{R_1}}$
16	zero
17	false
18	M
19	$\dfrac{1}{64}$
20	ellipses
21	none
22	false
23	true
24	Mg
25	$\dfrac{1}{8}$
26	false
27	false
28	true
29	true
30	1
31	$\dfrac{4\pi^2 R}{T^2}$

Item Number	Answer
32	$\dfrac{1}{G}$
33	$\dfrac{mv_1^2}{R_1}$
34	true
35	true
36	4
37	weight
38	$\dfrac{1}{8}$
39	$\sqrt{GM}$
40	false
41	A
42	$\dfrac{4\pi^2 R^3}{GT^2}$
43	gravitational attraction
44	5
45	$F' - F_g$
46	true
47	true
48	$\dfrac{1}{\sqrt{G}}$
49	5
50	double
51	4
52	true
53	true
54	$4\,Mg$
55	true
56	true
57	false
58	true
59	none
60	gravitational attraction
61	Kepler
62	$\dfrac{1}{r^2}$

Chapter 5 Answer Sheet

Item Number	Answer	Item Number	Answer	Item Number	Answer
1	false	48	false	100	true
2	$\frac{1}{4}MV_0^2$	49	2	101	Fx
		50	g	102	a
3	positive	51	4	103	$\frac{1}{2}V_0$
4	$2\,MgL$	52	clockwise		
5	LF	53	the same	104	$\frac{1}{4}MV_0^2$
6	counterclockwise	54	true		
7	$\frac{1}{2}mgR$	55	MgL	105	1
		56	true	106	b
8	b	57	true	107	true
9	1	58	LF	108	the same as
10	decreases	59	$4g$	109	zero
11	true	60	2		
12	true	61	false	110	$\frac{1}{2}MV_0^2 - 2\,Mgh'$
13	$\frac{1}{2}$	62	mgR	111	false
		63	true	112	$-c$
14	true	64	lost	113	decrease
15	true	65	r_1 to r_2	114	3
16	zero	66	5	115	true
17	$\dfrac{R_1}{R_2}$	67	mgR	116	5
		68	remain the same	117	true
18	increase	69	3	118	higher
19	true	70	zero		
20	false	71	$\dfrac{V_0^2}{8g}$	119	$\dfrac{V_0^2}{2g}$
21	conservative	72	true	120	$Mg\,\dfrac{\cos\theta}{2\sin\theta}$
22	F_h	73	false	121	MV_0
23	true	74	a	122	heat energy
24	$\frac{1}{2}Mg$	75	true	123	$\dfrac{M_1x_1' + M_2x_2' + M_3x_3'}{M}$
25	$(a+b)$	76	LF		
26	$\frac{1}{4}MV_0^2$	77	$\frac{1}{2}$	124	true
		78	neither	125	true
27	true	79	gm cm/sec^2	126	$\frac{1}{4}$
28	4	80	clockwise		
29	b	81	decrease	127	$GM_1M_2\left(\dfrac{1}{r_1} - \dfrac{1}{r_2}\right)$
30	$-a$	82	true		
31	a	83	in opposing	128	$\frac{1}{2}MV_0^2$
32	any point	84	false		
33	neither	85	$\dfrac{M_1x_1 + M_2x_2 + M_3x_3}{M_1 + M_2 + M_3}$	129	false
34	2			130	false
35	MgL	86	$-c$	131	$-xF$
36	c	87	2	132	greater
37	$(a+b)$	88	a proton	133	R_2F
38	$(2M)gh'$	89	c	134	$4g$
39	2 to 1	90	similar	135	$\frac{1}{2}MgL\cos\theta$
40	$\frac{1}{2}mV_0^2$	91	same		
		92	Mgh	136	true
41	$(L+x)F$	93	true	137	true
42	$MR_2V\perp$	94	true	138	true
43	less than	95	clockwise	139	negative
44	false	96	true	140	$\frac{1}{2}MV_0^2$
45	false	97	$\sqrt{4gL}$		
46	$-(a+b)$	98	$F_h L \sin\theta$	141	2
47	true	99	same as	142	external

Chapter 6 Answer Sheet

Item Number	Answer
1	$F\Delta L$
2	$1\frac{1}{2}$
3	2
4	3
5	all
6	$\frac{3}{2}N_0k$
7	false
8	true
9	$\frac{5}{3}$
10	energy per unit volume
11	$\frac{2}{3}$
12	neither
13	false
14	$\frac{3}{2}NkT$
15	$\frac{1}{T}$
16	100
17	true
18	true
19	M_1
20	true
21	less
22	greater than
23	$\frac{DV}{m}$
24	all
25	true
26	false
27	true
28	$\frac{M_1}{V_w}$
29	18
30	76
31	true
32	$\frac{5}{2}N_0k$
33	1
34	true
35	2
36	true

Item Number	Answer
37	increase
38	$\frac{3}{2}Nk$
39	true
40	Nm
41	lower than
42	$P\Delta V$
43	false
44	into
45	$V_0^{\frac{1}{3}}$
46	true
47	N_0
48	true
49	true
50	true
51	true
52	false
53	false
54	N_0k
55	10^6
56	remain the same
57	2
58	true
59	false
60	18
61	$\frac{M_1}{D_w}$
62	remain the same
63	false
64	true
65	$\frac{5}{2}$
66	3
67	$P\Delta V$
68	32
69	38
70	true
71	rise
72	true
73	NkT
74	$C_V + N_0k$
75	1
76	$\frac{M}{D_0}$
77	N_0k

Chapter 7 Answer Sheet

Item Number	Answer
1	ball
2	left
3	less than
4	4
5	$4\pi\sigma$
6	$\dfrac{e^2}{r_c^2}$
7	$r + \dfrac{L}{2}$
8	qEx
9	$4\pi(Q_1 + Q_2 + Q_3)$
10	3.01×10^{14}
11	false
12	D
13	neither
14	negative
15	9×10^9
16	1.53×10^{-16}
17	positive
18	zero
19	$\dfrac{-e^2}{2r_c}$
20	less
21	increase
22	$\dfrac{\text{kg meter}^3}{\text{sec}^2 \text{ coul}^2}$
23	attracted
24	repelled
25	Q_2
26	$-W_0$
27	$\dfrac{e^2}{r^2}$
28	zero
29	hanging together
30	newtons
31	decrease
32	14.4×10^4
33	true
34	attracted
35	true
36	1
37	true
38	4
39	$4\pi(x + R_3)^2$
40	$\sqrt{GM_p/r}$
41	decrease
42	$Q_1 E$
43	$KE - e^2/r$
44	positive
45	right
46	negative
47	true

Item Number	Answer
48	true
49	$\sqrt{Fr^2}$
50	positive
51	$-\rho_2 E_1$
52	9×10^9
53	left
54	false
55	$\dfrac{-e^2}{r_0}$
56	more
57	2
58	coulombs
59	heavier
60	9×10^9
61	$\dfrac{1}{q}\Delta U$
62	true
63	$\dfrac{GM_p m}{r^2}$
64	d'
65	150
66	Q
67	right
68	$KE + U$
69	leaves
70	r_c
71	zero
72	higher
73	7.4×10^3
74	true
75	5
76	negative
77	than
78	4.8×10^{-10}
79	negative
80	larger
81	electric force
82	opposite
83	electrons
84	leave
85	leaves
86	$\dfrac{-e^2}{r_0}$
87	9×10^{14}
88	neither
89	$\dfrac{10}{10^2}$
90	$\dfrac{4\pi Q}{A}$
91	down
92	$\dfrac{-\Delta U}{\Delta x}$

Item Number	Answer	Item Number	Answer
93	$\dfrac{Q}{R}$	126	qE
94	the same as	127	come together
95	$\dfrac{(3 \times 10^9)^2}{100^2}$	128	$(Q_1 + Q_2 + Q_3)$
		129	true
96	$\text{gm}^{\frac{1}{2}}\,\text{cm}^{\frac{3}{2}}\,\text{sec}^{-1}$	130	$\dfrac{E}{\epsilon}$
97	$d - d'$		
98	$\dfrac{Q_1 + Q_2 + Q_3}{(x + R_3)^2}$	131	$\dfrac{4\pi Q}{A}(d - d')$
		132	false
99	true	133	true
100	leaves	134	r_a to r_b
101	true	135	ball
102	$\dfrac{mv^2}{r}$	136	$\dfrac{100}{20^2}$
		137	qEx
103	true	138	$\dfrac{2\rho_1\rho_2}{R}$
104	out of		
105	positive	139	3×10^9
106	$\dfrac{Q}{(r + \tfrac{1}{2}L)^2}$	140	none
		141	repulsive
107	right	142	left
108	false	143	$4\pi(Q_1 + Q_2 + Q_3)$
109	positive	144	$\tfrac{1}{2}r_0$
110	$4\pi(Q_1 + Q_2)$		
111	the circumference by v	145	positive
112	$-\dfrac{1}{2}U$	146	$\dfrac{A}{4\pi\left(d - d' + \dfrac{d'}{\epsilon}\right)}$
113	3.01×10^{23}		
114	true	147	$\dfrac{\sigma(\epsilon - 1)}{\epsilon}$
115	$\dfrac{GM_p m}{r^2}$	148	2 to .1
116	$-qE\Delta x$	149	$\dfrac{Q}{(r - \tfrac{1}{2}L)^2} - \dfrac{Q}{(r + \tfrac{1}{2}L)^2}$
117	neither		
118	$\dfrac{1}{2}$	150	true
		151	true
119	$\dfrac{4\pi Q}{A}\left(d - d' + \dfrac{d'}{\epsilon}\right)$	152	q from ∞ to its present position
120	$4\pi(\sigma - \sigma')$	153	$\dfrac{2\rho_1}{R}$
121	$\dfrac{-\Delta V}{\Delta x}$	154	none
		155	positive
122	$\dfrac{Q}{R}$	156	$\dfrac{e^2}{2r_c}$
123	ball	157	$4\pi(\sigma - \sigma')$
124	true	158	positive
125	$\dfrac{e^2}{GM_p m}$	159	less
		160	true

Chapter 8 Answer Sheet

Item Number	Answer	Item Number	Answer	Item Number	Answer
1	part of a circle	45	$\dfrac{2I}{cR}$	89	qE
2	$\dfrac{2I}{cr}$	46	.02	90	zero
3	parallel	47	attractive	91	l and B
4	3×10^{10}	48	B	92	leave the region
5	right	49	parallel	93	neither
6	zero	50	$Nv_1 e$	94	in
7	II	51	$\dfrac{4\pi i L_1}{c}$	95	neither
8	$2BL_1$			96	cancel
9	field produced by a current	52	true	97	$V_0 - V'$
10	2 dynes	53	true	98	Ampere's
11	straight line	54	$2\pi R$	99	mBL
12	1	55	$Ne(v_1 + v_2)$	100	mB
13	neither	56	zero	101	clockwise
14	left	57	2×10^{-7}	102	would not
15	$\dfrac{evB}{c}$	58	$\dfrac{(Il)}{c} B \sin\theta$	103	amperes
				104	magnetic
16	right	59	remain the same	105	true
17	E	60	iL_1	106	the same as
18	zero	61	true	107	motor
19	$\dfrac{2I(l'l')}{c^2 r}$	62	zero	108	$\dfrac{N}{c} \dfrac{\Delta N_B}{\Delta t}$
		63	false	109	clockwise
20	nI	64	A	110	$\dfrac{4\pi i}{c}$
21	true	65	e		
22	2×10^{-7}	66	down	111	counterclockwise
23	$\dfrac{qvB}{c}$	67	lower	112	$B_r 2\pi r$
		68	left	113	true
24	B	69	d	114	true
25	Nev_2	70	will not	115	ac
26	opposite	71	true	116	statvolts
27	3×10^{10}	72	zero	117	enclosed current
28	true	73	along B	118	$2\pi r$
29	counterclockwise	74	$\dfrac{Ia_1 B}{c}$	119	$\dfrac{iA}{c}$
30	Ampere's	75	along B	120	more negative than
31	$\dfrac{l}{10}$	76	$\dfrac{4\pi i}{c}$	121	parallel
32	right	77	$\dfrac{1}{r}$	122	$\dfrac{2\pi i}{c}$
33	in			123	erg/statcoul
34	left	78	BL_1	124	$\dfrac{NIBA}{c}$
35	$\dfrac{4\pi i}{c}$	79	true		
		80	none	125	electric
36	$\dfrac{2I}{cr}$	81	right	126	perpendicular
		82	true	127	$\dfrac{\Delta I}{\Delta t}$
37	$\dfrac{qv}{c} B \sin\theta$	83	true		
		84	$\dfrac{2I}{cr}$	128	upper
38	true	85	$\dfrac{1}{r}$	129	$\dfrac{BvL_3}{c}$
39	along the circle				
40	right	86	B	130	infinite
41	down	87	$\dfrac{.2}{100}$	131	would
42	c			132	true
43	down	88	B	133	$\Sigma E_i \cdot L_i$
44	force on a current in a field			134	surface current

Item Number	Answer	Item Number	Answer	Item Number	Answer
135	true	163	B	188	in
136	$\frac{1}{2}F_1 a_2$	164	$\frac{r}{2c}\left(\frac{\Delta B}{\Delta t}\right)$	189	$2\pi rE$
137	Faraday's			190	18
138	potential	165	EL_3	191	generator
139	true	166	$\frac{NI}{L}$	192	$\frac{r}{2c}\left(\frac{\Delta E}{\Delta t}\right)$
140	FL	167	true	193	E
141	$\frac{\pi r^2}{c}\left(\frac{\Delta E}{\Delta t}\right)$	168	Maxwell's modification	194	$EB = \frac{EBv^2}{c^2}$
142	along B	169	$\frac{2I}{cr}$	195	statvolts
143	true	170	v	196	true
144	$\frac{4\pi N^2 A}{c^2 l}$	171	$\frac{iBAL}{c}$	197	BL_3
145	$B \cdot A$	172	gauss	198	$\Sigma E_i \cdot l_i = \frac{1}{c}\left(\frac{\Delta N_B}{\Delta t}\right)$
146	mB	173	$\frac{4\pi I}{c}$	199	true
147	perpendicular			200	$\Sigma B_i \cdot l_i = \frac{4\pi I}{c}$
148	$\frac{4\pi N^2 A}{c^2 l}$	174	$\frac{1}{c}\left(\frac{\Delta N_B}{\Delta t}\right)$	201	will
149	left end	175	a	202	horizontal
150	counterclockwise	176	will not	203	statamps
151	$\frac{evB}{c}$	177	zero	204	Right hand rule I
		178	left	205	1
152	opposed to B	179	6	206	true
153	$F_1 a_2$	180	$B_L \cdot L = \frac{4\pi I}{c} + \frac{1}{c}\left(\frac{\Delta N_E}{\Delta t}\right)$	207	$\Sigma B_i \cdot l_i = \frac{4\pi I}{c} + \frac{1}{c}\left(\frac{\Delta N_E}{\Delta t}\right)$
154	$\frac{4\pi NA}{cl}\frac{\Delta I}{\Delta t}$	181	magnetic	208	6
155	$\frac{Ir^2}{R^2}$	182	$\frac{1}{6} \times 10^{-5}$	209	$BL_3 x$
156	$B\pi r^2$			210	Ampere's law
157	true	183	$E = \frac{Bv}{c}$	211	true
158	EL_3	184	true	212	$\Sigma E_i \cdot l_i$
159	true	185	.02	213	different from
160	$2\pi Br$	186	zero	214	Ampere's law, modified
161	iL_1	187	$\frac{4\pi I}{c}\left(\frac{N}{l}\right)$	215	$\frac{4\pi I}{c}$
162	$\frac{N}{c}\frac{\Delta N_B}{\Delta t}$				

Chapter 9 Answer Sheet

Item Number	Answer
1	ergs
2	decrease
3	parallel
4	3×10^2
5	same
6	infinite
7	I^2R
8	3
9	$\dfrac{2\pi}{T}$
10	2
11	false
12	2
13	$\eta e \dfrac{\Delta u}{2}$
14	$\dfrac{V}{L}$
15	8×10^{-4}
16	c
17	$\dfrac{1}{\dfrac{1}{R_1}+\dfrac{1}{R_2}+\dfrac{1}{R_3}}$
18	382
19	c
20	2
21	inversely
22	6
23	5
24	$\dfrac{V^2}{R}$
25	V_1
26	a
27	true
28	increases
29	directly
30	6
31	$\dfrac{V}{R}$
32	2
33	true
34	$\eta A v_d$
35	2
36	1
37	parallel
38	halved
39	b
40	R_1
41	zero
42	d
43	volts
44	true

Item Number	Answer
45	$\dfrac{\eta e^2 L}{2mu}$
46	$\dfrac{10^{-6} \times 3 \times 10^9}{4.8 \times 10^{-10}}$
47	iA
48	zero
49	minimum
50	zero
51	decreased
52	d
53	conducting
54	true
55	true
56	$\dfrac{R_3R_2 + R_1R_3 + R_2R_1}{R_1R_2R_3}$
57	a
58	zero
59	true
60	white
61	zero
62	neither
63	20 cm
64	$\dfrac{EL}{iA}$
65	no
66	X
67	500
68	forward
69	eB
70	$\dfrac{i}{E}$
71	true
72	5×10^8
73	1 erg
74	$\dfrac{evB}{c}$
75	$\dfrac{V}{I}$
76	$I_1 - I_3$
77	A
78	2
79	240 watts
80	$\dfrac{L}{\sigma A}$
81	gray
82	5
83	8
84	3
85	$\eta e \bar{v}_d$
86	$\eta A e$
87	6.1×10^{-4}
88	true

Item Number	Answer
89	magnetic force
90	$\dfrac{\Delta u}{2}$
91	amperes
92	a
93	$-V_2$
94	zero
95	false
96	1 joule
97	2
98	381.5
99	$\dfrac{R_1 R_2}{R_1 + R_2}$
100	$\dfrac{400}{\pi}$
101	false
102	$\eta A \bar{v}_d e$
103	4 volts
104	IR
105	6
106	true
107	$\dfrac{10}{3 \times 10^{10}}$
108	zero

Chapter 10 Answer Sheet

Item Number	Answer
1	$D_3 - D_4 + d\sin\theta$
2	true
3	violet
4	$\dfrac{v}{f}$
5	$\left(N + \dfrac{1}{2}\right)\lambda$
6	true
7	true
8	$\dfrac{1}{f}$
9	$D_1 + D_3 - D_2 - D_4$
10	true
11	false
12	true
13	$\dfrac{2}{T}$
14	$\dfrac{1}{2}\lambda$
15	true
16	$\dfrac{\lambda}{c}$
17	$\dfrac{1}{2}$
18	true
19	true
20	$\dfrac{f}{c}$
21	square
22	$\dfrac{1}{4}$
23	reflection
24	pass through C
25	$\dfrac{1}{2}ky_0^2$
26	$\dfrac{1}{n}$
27	PO
28	$\dfrac{t}{T}$
29	3
30	less
31	2
32	diverging
33	never
34	larger
35	$\dfrac{d}{N}$
36	increases
37	$\dfrac{1}{4}$
38	true
39	1.43×10^4
40	smaller
41	$N\lambda$
42	$\dfrac{c}{\lambda}$
43	increases
44	$A'B'O$
45	$\left(N + \dfrac{1}{2}\right)\lambda$
46	minimum
47	greater
48	$\dfrac{1}{T}$
49	true
50	$\dfrac{N\lambda + D_4 - D_3}{d}$
51	true
52	increase
53	neither
54	2
55	true
56	3.75×10^9
57	none
59	$\dfrac{1}{2}$
60	be parallel to CB
61	0.667
62	true
63	$\dfrac{1}{2}ky_0$
64	$A'B'F$
65	true
66	square
67	true
68	negative
69	true
70	$\dfrac{s'}{s}$
71	true
72	$\dfrac{1}{s} + \dfrac{1}{s'}$
73	D
74	image
75	false
76	6 inches
77	$\dfrac{\lambda}{d}$
78	$\dfrac{s' - f}{f}$
79	true
80	$\dfrac{N\lambda}{d}$
81	twice
82	$\dfrac{1}{2}\lambda$
83	true
84	$\dfrac{1}{2}$
85	between P and F
86	$\dfrac{1}{2}$
87	8
88	c
89	$\left(N + \dfrac{1}{2}\right)\lambda$
90	d
91	$\dfrac{\lambda}{c}$
92	maximum
93	$\dfrac{\sin\theta_1}{\sin\theta_2} = \dfrac{1}{n}$
94	$N\lambda$
95	F
96	$\dfrac{1}{2}$
97	false
98	S_5
99	true
100	b
101	twice
102	H
103	right
104	$D_1 + D_3 - D_2 - D_4 = \left(N + \dfrac{1}{2}\right)\lambda$
105	$\dfrac{120}{n}$
106	$\dfrac{1}{2}\lambda$
107	above
108	true
109	greater
110	true
111	F
112	c
113	yes
114	vT
115	24 inches in front of the mirror

Chapter 11 Answer Sheet

Item Number	Answer
1	$10\sqrt{3}$
2	B
3	$\dfrac{1.7c}{1.72}$
4	left
5	true
6	neither
7	β
8	Mv
9	slow
10	$\gamma(x' + vt')$
11	L'
12	0.6
13	greater
14	$\gamma\left(t - \dfrac{vx}{c^2}\right)$
15	u'
16	shorter
17	$M_b(v_b)_y$
18	6
19	ct'
20	Δt
21	t
22	running slow by a factor of 0.6
23	true
24	2
25	$\dfrac{0.1c}{0.28}$
26	$\dfrac{u - v}{1 - vu/c^2}$
27	6
28	7
29	moving
30	γ
31	0.6
32	1
33	greater
34	$\dfrac{1}{2}\beta^2 M_0 c^2$
35	60 sec
36	true
37	1.33
38	more
39	$\left(1 - \dfrac{v^2}{c^2}\right)^{-\frac{1}{2}}$
40	neither
41	faster
42	B
43	left
44	$\dfrac{1}{\sqrt{1 - (v/c)^2}}$
45	true

Item Number	Answer
46	vt
47	1.67
48	$p^2 c^2$
49	$\gamma\left(ct' + \dfrac{vx'}{c}\right)$
50	true
51	true
52	$-vt'$
53	B
54	γ
55	$\gamma(x - vt)$
56	1
57	faster
58	A
59	more
60	slower
61	greater
62	less
63	B
64	true
65	$M_0 \gamma$
66	shorter
67	1
68	boxcar
69	true
70	$1.67 M_0 v$
71	$\dfrac{M_a}{\sqrt{1 - (v/c)^2}}$
72	6
73	15
74	$\gamma\beta$
75	6
76	L
77	γt
78	less
79	2
80	true
81	contracted
82	Mc^2
83	$(v'_b)_y$
84	10
85	$\dfrac{u + v}{1 + vu/c^2}$
86	W^2
87	slower
88	false
89	$\gamma(x_2 - vt)$
90	true
91	M_b
92	$\gamma(x_2 - x_1)$
93	contracted
94	zero
95	true

Item Number	Answer	Item Number	Answer
96	true	137	$\dfrac{-GMm}{R}$
97	1/sec	138	true
98	$0.67\, M_0 c^2$	139	false
99	$\Delta t'$	140	$\dfrac{D^2}{\sigma}$
100	$\dfrac{0.1c}{0.28}$	141	25
101	none	142	(b)
102	0.8	143	true
103	true	144	true
104	same as	145	$\dfrac{L'}{\gamma}$
105	false		
106	away from	146	$1 - \dfrac{1}{\gamma}$
107	true	147	away from
108	true	148	$R\sqrt{\dfrac{8}{3}\pi GD}$
109	true		
110	L'	149	$\dfrac{L'}{\gamma}$
111	$\gamma(x + vt)$		
112	smaller	150	$\gamma\beta$
113	left	151	true
114	$(M - M_0)c^2$	152	$K(x_2 - x_1)^n$
115	$\dfrac{D^2}{\sigma}$	153	$\dfrac{D^2}{\sigma}$
116	$\gamma - 1$	154	$\gamma\beta M_0 c^2$
117	$K(x_2^n - x_1^n)$		
118	true	155	$\dfrac{c}{K}$
119	true	156	left
120	true	157	$\dfrac{\beta}{c}$
121	true		
122	$-GDmR^2$	158	1
123	$\gamma\left(t + \dfrac{vx}{c^2}\right)$	159	10^{17}
		160	1
124	$\dfrac{1.7c}{1.72}$	161	(b)
		162	A
125	true	163	true
126	$\gamma M_0 c^2$	164	β^2
127	true	165	steady state
128	$\dfrac{-u + v}{1 - vu/c^2}$	166	first
129	recession	167	$\dfrac{4}{3}\pi R^3 D$
130	galaxy		
131	true	168	contracted
132	true	169	$M_0^2 c^4$
133	false		
134	$K x_1^n$		
135	β		
136	$\gamma\beta M_0 c$		

Chapter 12 Answer Sheet

Item Number	Answer	Item Number	Answer	Item Number	Answer
1	increases	45	$\dfrac{h}{\sqrt{2mKE}}$	87	c
2	2	46	true	88	true
3	2	47	2	89	c
4	e	48	decreases	90	$hf - KE_{max}$
5	true	49	$\dfrac{nA}{N_a N}$	91	2
6	9	50	5	92	more than 90°
7	$\dfrac{1}{2000}$	51	true	93	3 times
8	minimum	52	A	94	number
9	false	53	$\dfrac{2KE}{q}\sqrt{\dfrac{\sigma}{\pi}}$	95	3085Å
10	$\dfrac{hf}{c}$	54	atoms	96	false
11	5	55	true	97	2
12	energy	56	b	98	true
13	false	57	$\dfrac{W}{c^2}$	99	true
14	$\sqrt{3}$	58	negative	100	false
15	$KE_0 + w$	59	19	101	3
16	$\dfrac{P^2}{2m}$	60	photons	102	b
17	false	61	more	103	2.7^2
18	3σ	62	5	104	$\dfrac{hc}{W}$
19	electrostatic	63	d	105	19^2
20	atoms in the foil	64	true	106	true
21	zero	65	4	107	3
22	$\dfrac{1}{4000}$	66	true	108	true
23	$N_a \dfrac{3\sigma}{A}$	67	Ze	109	$hf - w$
24	9	68	12		
25	a	69	b		
26	Coulomb's	70	$\sqrt{2mKE}$		
27	false	71	true		
28	$\dfrac{hc}{W}$	72	$\pi\left(\dfrac{Qq}{mv^2}\right)^2$		
29	c	73	false		
30	true	74	true		
31	horizontal	75	9		
32	-5	76	$\dfrac{3\sigma}{A}$		
33	$\dfrac{N\sigma}{A}$	77	361		
34	1	78	$\dfrac{4(KE)^2}{\pi q^2}$		
35	alpha particles	79	intensity		
36	hf	80	true		
37	$\dfrac{\frac{1}{4}\pi Q^2 q^2}{(KE)^2}$	81	$\dfrac{Qq}{mv^2}$		
38	2	82	a, e		
39	-5	83	1.8×10^{-32}		
40	$\dfrac{N\sigma}{A}$	84	$\dfrac{h^2}{2m\lambda^2}$		
41	greater than b	85	$W - w$		
42	d	86	$\dfrac{N_a N\sigma}{A}$		
43	7				
44	none				

Chapter 13 Answer Sheet

Item Number	Answer	Item Number	Answer	Item Number	Answer
1	$2R$	45	1	90	true
2	1	46	1	91	centripetal force
3	more	47	$2R$	92	13.6
4	$\dfrac{4R}{N}$	48	13.6	93	$\dfrac{h}{\lambda}$
5	true	49	wavelength	94	Bohr
6	(c)	50	$2(2l+1)$	95	true
7	6	51	Schroedinger's	96	10
8	2	52	zero	97	$-\dfrac{1}{2}$
9	2	53	$\dfrac{h}{p}$	98	true
10	4	54	$\dfrac{Ze^2}{R^2}$	99	8
11	6	55	1	100	$\dfrac{2L}{N}$
12	true	56	true		
13	N^2	57	greater	101	$\dfrac{h}{\sqrt{\dfrac{mZe^2}{R_{av}}}}$
14	2.8	58	(a)		
15	4	59	10	102	λ
16	-13.6	60	2	103	14
17	$4R$	61	false	104	$-l$
18	1.25	62	2	105	$-\dfrac{2\pi^2 mZ^2 e^4}{N^2 h^2}$
19	zero	63	$\dfrac{N^2 h^2}{16 mZe^2}$	106	1
20	true	64	$2l+1$	107	6
21	$\dfrac{1}{2}$	65	15	108	0
22	position	66	lower	109	2
23	2	67	2		
24	$\dfrac{1}{9}$	68	1		
25	b	69	b		
26	8	70	2		
27	false	71	-13.6		
28	2	72	2		
29	$\dfrac{h}{\sqrt{2mKE}}$	73	$\dfrac{h}{\lambda}$		
30	$\dfrac{1}{2}$	74	right		
31	false	75	$(a+b)$		
32	lower	76	12		
33	13.6	77	false		
34	true	78	-1		
35	2	79	2		
36	$\dfrac{1}{2}N$	80	1		
37	4, 0	81	0.5		
38	angular momentum	82	A		
39	2	83	1		
40	$\dfrac{hf}{c}$	84	$\dfrac{1}{2}$		
41	6	85	N		
42	true	86	19		
43	smaller	87	8		
44	$-\dfrac{1}{2}$	88	$\dfrac{h^2}{16 mZe^2}$		
		89	less		

Chapter 14 Answer Sheet

Item Number	Answer	Item Number	Answer
1	5	43	zero
2	work functions	44	2
3	A to B	45	3
4	true	46	true
5	30,000	47	1
6	$hf - W$	48	1
7	molecule	49	false
8	2	50	true
9	3	51	$\dfrac{3}{2}$
10	release	52	5
11	2	53	neither
12	1	54	$a - 4kT$
13	2	55	2
14	1	56	negative
15	-5	57	maximum kinetic
16	1	58	B
17	b	59	negative
18	true	60	true
19	false	61	7
20	$\dfrac{1}{256}$	62	$\dfrac{1}{4}$
21	covalent	63	negative
22	A	64	$hf - KE_{max}$
23	$KE_0 + W$	65	square
24	true	66	(b)
25	2 ev	67	5
26	less	68	4.48
27	8 ev	69	reduced
28	up	70	$\dfrac{h^2}{2m\lambda^2}$
29	2	71	true
30	divide	72	C
31	$\dfrac{1}{16}$	73	$4.48 \times 1.6 \times 10^{-12}$
32	10^{-2}	74	2
33	away from	75	$\dfrac{1}{64}$
34	work functions	76	increased
35	increased	77	true
36	$2 \times 13.6 + 4.48$	78	false
37	$15 - 13.6$	79	kinetic
38	true	80	increases to infinity
39	a		
40	4.3×10^{12}		
41	B		
42	2		

Chapter 15 Answer Sheet

Item Number	Answer
1	true
2	true
3	A
4	$\frac{1}{4}\lambda$
5	$\frac{1}{\sqrt[3]{2}}$
6	He^4
7	$\frac{N_a \sigma}{a}$
8	decrease
9	true
10	false
11	true
12	B
13	3.0×10^{-13}
14	$\frac{1}{2}$
15	29
16	$\frac{1}{2}$
17	1
18	$\frac{(N-N')a}{NN_a}$
19	true
20	greater
21	$2^{\frac{1}{4}}$
22	true
23	true
24	50
25	tritium
26	increase
27	$\frac{M}{A}$
28	true
29	20
30	$(N')^2$
31	5000
32	less than
33	mass number
34	6.25
35	true
36	20
37	$\frac{83}{30 \times 1.6 \times 10^{-12}}$
38	true
39	1.6×10^{-4}
40	$\frac{1}{4}$
41	false

Item Number	Answer
42	neutrons
43	increase
44	70
45	3×10^6
46	$\frac{NN_a \sigma}{a}$
47	heavier
48	weaker
49	true
50	Fe
51	10
52	0.2
53	6
54	true
55	$\frac{1}{2}$
56	.52
57	true
58	10
59	R
60	20
61	29
62	true
63	0.2
64	true
65	true
66	true
67	4×10^{-4}
68	10
69	830
70	10
71	$\frac{1}{2}$
72	3
73	increase
74	10
75	4.0
76	true
77	4.8×10^{-10}
78	true
79	36
80	250
81	plus
82	false
83	false
84	930
85	10
86	110
87	200
88	2×10^{-8}
89	45
90	2

Chapter 16 Answer Sheet

Item Number	Answer	Item Number	Answer	Item Number	Answer
1	1	56	false	110	weak
2	1	57	cannot	111	nuclear
3	true	58	-1	112	yes
4	charge	59	meson	113	$\bar{\nu}_\mu$
5	-1	60	strong	114	electromagnetic
6	lepton	61	both sides	115	counterclockwise
7	false	62	no	116	yes
8	γ, ν, e, P	63	hyperon		
9	energy	64	left		
10	permitted	65	ν_e		
11	true	66	true		
12	permitted	67	ν, e, μ		
13	same	68	weak		
14	-1	69	can		
15	positive	70	false		
16	heavy particles	71	cannot		
17	forbidden	72	no		
18	-1	73	2		
19	right	74	-1		
20	anti-lambda	75	true		
21	true	76	leptons		
22	forbidden	77	opposite		
23	heavy particles	78	electromagnetic		
24	true	79	1		
25	meson	80	antiparallel		
26	-1	81	can		
27	ν_e	82	ν_μ		
28	right	83	weak		
29	hyperon	84	1		
30	neutrons	85	2		
31	2	86	true		
32	$\pi, K, P, N, \Lambda, \Sigma, \Xi, \Omega$	87	yes		
33	true	88	counterclockwise		
34	hyperon	89	strong		
35	north	90	weak		
36	can	91	true		
37	charge	92	antineutrino		
38	$K, \Lambda, \Sigma, \Xi, \Omega$	93	false		
39	strong	94	cannot		
40	forbidden	95	permitted		
41	leptons	96	$P, N, \Lambda, \Sigma, \Xi, \Omega$		
42	true	97	nuclear		
43	weak	98	3		
44	lepton	99	against		
45	1	100	same		
46	true	101	true		
47	positive	102	up		
48	ν	103	nuclear		
49	0	104	antiparallel		
50	true	105	false		
51	true	106	yes		
52	electromagnetic	107	left		
53	antineutrino	108	true		
54	lepton	109	up		
55	$\bar{\nu}_e$				